Radiative Heat Transfer
by the Monte Carlo Method

ADVANCES IN HEAT TRANSFER

Volume 27

Radiative Heat Transfer by the Monte Carlo Method

Advances in
HEAT TRANSFER

Wen-Jei Yang
Department of Mechanical
  Engineering and
  Applied Mechanics
University of Michigan
Ann Arbor, Michigan

Hiroshi Taniguchi
Faculty of Engineering
Hokkaido University
Chuo-ku
Sapporo 064
Japan

Kazuhiko Kudo
Faculty of Engineering
Hokkaido University
Sapporo 064
Japan

Serial Editors

James P. Hartnett
Energy Resources Center
University of Illinois
Chicago, Illinois

Thomas F. Irvine
Department of Mechanical Engineering
State University of New York at Stony Brook
Stony Brook, New York

Serial Associate Editors

Young I. Cho
Department of Mechanical
  Engineering
Drexel University
Philadelphia, Pennsylvania

George A. Greene
Department of Advanced Technology
Brookhaven National Laboratory
Upton, New York

Volume 27

ACADEMIC PRESS
San Diego Boston New York London Sydney Tokyo Toronto

This book is printed on acid-free paper. ∞

Copyright © 1995 by ACADEMIC PRESS, INC.

All Rights Reserved.
No part of this publication may be reproduced or transmitted in any form or by any means, electronic or mechanical, including photocopy, recording, or any information storage and retrieval system, without permission in writing from the publisher.

Academic Press, Inc.
A Division of Harcourt Brace & Company
525 B Street, Suite 1900, San Diego, California 92101-4495

United Kingdom Edition published by
Academic Press Limited
24-28 Oval Road, London NW1 7DX

International Standard Serial Number: 0065-2717

International Standard Book Number: 0-12-020027-9

PRINTED IN THE UNITED STATES OF AMERICA
95 96 97 98 99 00 QW 9 8 7 6 5 4 3 2 1

CONTENTS

Preface vii

Part I
PRINCIPLES OF RADIATION

1. Thermal Radiation

1.1 Introduction 3
1.2 Definitions and Laws Regarding Thermal Radiation 7
1.3 Definitions and Laws Regarding Gas Radiation 13
1.4 Definitions and Laws Regarding Radiation from
    Gas–Particle Mixtures 17

2. Radiation Heat Transfer

2.1 Basic Equations 19
2.2 Existing Methods of Solutions 23

Part II
PRINCIPLES OF MONTE CARLO METHODS

3. Formulation

3.1 Introduction 45
3.2 Heat Balance Equations for Gas Volumes and Solid Walls . . 46
3.3 Simulation of Radiative Heat Transfer 49

4. Methods of Solution

4.1 Energy Method 86
4.2 READ Method 86

5. Special Treatises

5.1 Introduction 92
5.2 Scattering by Particles 93
5.3 Nonorthogonal Boundary Cases 99

Part III
APPLICATIONS OF THE MONTE CARLO METHOD

6. Two-Dimensional Systems

6.1	Introduction	107
6.2	Radiative Heat Transfer in Absorbing–Emitting Gas: Program RADIAN	107
6.3	Radiative Heat Transfer between Surfaces Separated by Nonparticipating Gas: Program RADIANW	130
6.4	Radiative Heat Transfer in Absorbing–Emitting and Scattering Media	146

7. Some Industrial Applications

7.1	Introduction	158
7.2	Boiler Furnaces	158
7.3	Gas Reformer	167
7.4	Combustion Chambers of Jet Engines	173
7.5	Nongray Gas (Combustion Gas) Layer	181
7.6	Circulating Fluidized Bed Boiler Furnace	187
7.7	Three-Dimensional Systems	193

References	200
Applications on Disk	203
List of Variables in Computer Programs	204
Author Index	209
Subject Index	211

PREFACE

Physical phenomena result from a combination of multiple basic processes. Often one can easily comprehend seemingly complex phenomena by investigating each of the basic processes. For example, the properties of a gas or liquid result from the interactions of the molecules which constitute the fluid, as described in molecular gas dynamics. The traffic flow in a highway network can be determined by summing the movement of individual automobiles. Likewise, radiative heat transfer treated in this monograph can be described as a summation of the behavior of individual energy particles. In other words, the radiative energy emitted from a body in proportion to the fourth power of local surface temperature is equivalent to the emission of multiple energy particles.

The traditional approach to these kinds of physical problems is to model the phenomena with mathematical equations and then solve these equations. However, this monograph adopts the Monte Carlo approach: Macroscopic physical phenomena are divided into a number of basic processes. The behavior of the individual or basic processes which is stochastic is then investigated. The behaviors of the individual processes in sum simulate the behavior of the entire physical phenomenon. Hence, two factors must be incorporated into the Monte Carlo method: the probability distribution for the occurrence of each basic process and the physical laws which these processes must obey. In general, a physical phenomenon is simplified upon its decomposition into basic processes, making it easier to consider the effects of various parameters and conditions. In contrast, it is generally difficult to take all affecting parameters and conditions into account in the analysis of a macroscopic physical phenomenon. Even if it is possible to include all the relevant parameters and conditions, it is difficult to obtain the solution of the resulting formulation. Hence, the modeling of macroscopic physical phenomena is limited to simpler cases.

Superiority of the Monte Carlo Method

In the analysis of radiative heat transfer, the conventional flux and zone methods cannot treat the problem of specular-reflection walls. The flux method can treat the problem of gas scattering with the aid of some bold assumptions, while the zone method cannot. Considerable effort is required to formulate multiple dimensional systems with complex boundary conditions, such as writing the governing equations corresponding to individual problems and the necessary boundary conditions. In comparison, the Monte Carlo method has flexibility in dealing with various parameters and conditions, for example, three dimensionality, arbitrary boundary geometry, arbitrary wall boundary conditions (diffuse reflection, specular reflection, specified heat flux, specified temperature), isotropic or anisotropic scattering, nonhomogeneity of physical properties, and a nongray body with wave length-dependent physical properties. In other words, it is possible to take into account all parameters and conditions which are actually encountered in radiation heat transfer. Another important merit of the Monte Carlo method is in the construction of computer programs. It is not necessary to rewrite the governing equations appropriate to each problem. One supplies the boundary geometry and physical properties as the input data, using the same program for analyzing radiative heat transfer. Even a novice in radiation theory can master, in a short time, the method of constructing computer programs for radiative heat transfer analysis by means of the Monte Carlo method. It is similar to the problem of traffic control. To formulate equations for the traffic quantity in a city traffic network requires much expertise, but everyone knows how to drive a car under city street conditions. Likewise, the Monte Carlo method decomposes a complex physical phenomenon into basic processes which can be treated using simple physical laws.

Research

Since 1963, we have been concerned with the practical applications of radiation heat transfer analysis using the Monte Carlo method. Research began with the analysis of gas absorption–scattering characteristics in a boiler. Since then, results were obtained for radiation analysis [1] in a one-dimensional system having internal heat generation and temperature-dependent physical properties, for radiation analysis [2] in a three-dimensional furnace of cubic geometry, and for temperature distribution [3] in the cubic furnace including the effect of convection. These theoretical results were compared with measurements [4] taken in an existing oil-burning boiler. It was demonstrated through these studies that both the gas temperature distribution and the wall heat-flux distribution can be determined if the distribution of radiative properties of gases and walls and the heat generation rate in the flame are known a priori. In the case of a combined radiation–convection heat transfer in duct flows between two parallel walls, the common practice was to employ a one-dimensional approximation considering only radiation perpendicular to the direction of flow. A study [5] was conducted to investigate how accurate the results obtained using the one-dimensional approximation would be if the problem were treated as two-dimensional including both the entrance and exit effects.

In order to enhance the practical uses of radiative heat transfer analysis by means of the Monte Carlo method, the Radiant Energy Absorption Distribution (READ) [6, 7] method was developed to reduce the computational time. This method had made it possible to simplify various practical problems having complex three-dimensional geometries and physical property distributions. Examples include the effect of flame shape on temperature uniformity of steel in a forge furnace [8], analysis of radiation from a combustion chamber to high-pressure turbine nozzle vanes in a jet engine [9], and analysis of radiative heat input to various parts of the human body in a floor-heated conference room with windows, tables, and chairs [10].

Simultaneously, studies were conducted to promote the application of the Monte Carlo method to gas–particle enclosures characterized by anisotropic scattering [11] and to compare the results with Menguc and Viskanta's analytical result [12] for a one-dimensional system for validation of the Monte Carlo method [13]. The method was extended to two-dimensional analyses of combined radiation–convection heat transfer in coal combustion boilers [14] and circulating fluidized beds [15] with absorbing–emitting–scattering gases. The latter case treated the mixed-phase flows of three different heat transfer media: heat-generating fuel particles, non-heat-generating bed particles, and combustion gas. Temperature differences between the combustion gas and the particles can be determined.

The continuum approximation method [16] is applied for radiation analysis of systems containing scattering particles with a postulation of uniform absorption, radiation, and scattering in the medium. However, it is known [17, 18] that the results obtained from this method begin to deviate from test data when the volume concentration of particles exceeds 10% (called a packed layer). In order to dissolve the limitation of this approximation method, we applied the Monte Carlo method to numerous experiments concerning the transmittance of radiant heat through regularly or irregularly packed spheres. At present, this result is used to investigate an extension of the continuous approximation method to treat a packed layer of up to 0.6 packing density.

Efforts are currently being directed to the application of the Monte Carlo method to the radiation analysis of nongray gases [19] and the transmission analysis of radiant energy through fibrous layers.

Applications of Monte Carlo Method in Industry

The following are examples of the application of the Monte Carlo method to radiation analyses in Japan. The largest number of applications may be found in combined radiation–convection heat transfer analyses in boiler furnaces, including an oil-fired boiler and a circulating fluidized bed boiler (Babcock Hitachi Company), a garbage incinerator (Mitsubishi Heavy Industries), an industrial oil-fired boiler (Takuma), and a marine boiler (Hitachi Shipbuilding). The second largest number of applications of the Monte Carlo method may be found in combined radiation–convection heat transfer analyses of various high-temperature heating furnaces and combustion chambers, including a forge furnace and a gas reformer (Tokyo Gas Company) and jet engine combustion chambers (Ishikawajima–Harima Heavy Industries). Other examples include a droplet radiator (a high-performance radiator for space stations; Ishikawajima–Harima Heavy Industries) and analysis of the floor heating process in these systems, which are characterized by a high-temperature field, a terrestrial environment, or a lack of forced-convective heat removal. An accurate evaluation of radiation heat transfer plays an important role in product design.

This monograph covers multidimensional, combined radiation–convection heat transfer in gray gases enclosed by gray walls. It consists of three parts. Part I presents the natural laws, definitions, and basic equations pertinent to radiative heat transfer and conventional methods for solving radiation heat transfer problems. Part II introduces the fundamentals of the Monte Carlo method. Energy balance equations are presented, followed by the simulation of radiative heat transfer, the procedure for determining temperature distribution, and treatises on scattering media and nonorthogonal boundaries. Part III presents applications of the Monte Carlo method. Examples include boiler furnaces, gas reformers, combustion chambers of jet engines, and circulating fluidized bed boilers. Numerous FORTRAN programs accompany the example problems to aid in understanding the application of the Monte Carlo method to computer programming.

Part I
PRINCIPLES OF RADIATION

Chapter 1
Thermal Radiation

1.1. Introduction

What is radiation? A simple answer is that radiation possesses dual characteristics of both electromagnetic waves and photons, the latter being particles having the smallest unit of energy, called quanta.

What are electromagnetic waves? Among several forms of them are the waves of radio, of light, and of X-rays. Of these three forms, radio waves have the greatest wavelengths, the wavelengths of light are intermediate, and those of X-rays are the shortest. Only light has thermal, i.e., heating, effects, which are of interest to radiation heat transfer, and thus light waves are the subject of interest here. Note that light and radiation will be used interchangeably in this section.

The basic nature of light may be considered under three categories:

1. The interaction of light with matter, with specific reference to the individual processes of the emission and absorption of light by atoms and molecules
2. The propagation of light through space and material media, which reveals the electromagnetic wave nature of light
3. The unification of the knowledge under categories 1 and 2.

The last category falls into the realm of the so-called "wave mechanics." In this field, it is demonstrated that the statistical average result of a large number of individual processes of emission, under category 1, leads to the nature that light is found to have in its propagation, under category 2. In other words, the unification is a statistical conclusion.

1.1.1. ELECTROMAGNETIC WAVES

In the year 1845, Faraday discovered that a beam of plane polarized light is rotated when passed through a bar of heavy flint glass in a magnetic field, thus establishing a relationship of light with electricity and magnetism. Twenty years later, in 1865, James Maxwell derived a set of differential equations for electricity and magnetisms. These differential equations had the same form as those for elastic and other mechanical waves. He conjectured that such a thing as electromagnetic waves might exist. In the year 1887, Heinrich Mertz found a way of generating a new form of these waves by using an electric oscillator as a source.

When electromagnetic waves are transmitted through empty space—regardless of the type or length—they have the same velocity in vacuum, that is, $c_0 = 2.99776 \times 10^8$ m/sec. The speed of light in a medium c is less than c_0 and is commonly given in terms of the index of refraction $n = c_0/c$, where n is greater than unity. For gases, n is very close to 1.

Let λ denote wavelength; τ, period; ν, frequency, and c, wave velocity. One has, for any train of waves,

$$\lambda = \nu\tau, \qquad (1.1a)$$

$$\nu = 1/\tau, \qquad (1.1b)$$

$$\nu\lambda = c, \qquad (1.1c)$$

where $c = 3 \times 10^8$ m/sec for electromagnetic waves in vacuum or air.

1.1.2. QUANTA

The emission and absorption of light occur in discrete packets or bundles of energy called *quanta*. Each of these bundles has a definite magnitude, but they do not have the same magnitude. Aggregates of quanta in their propagation unveil the characteristics of wave motion. That is, light in its propagation have a wave nature. Quantum theory deals with the laws of emission and absorption of light. Basic to this theory is the relation between the energy difference ΔE, between two stationary energy states, of levels E_n and E_{n+1}, for an atom or molecules, and the frequency ν and wavelength in vacuum λ of the electromagnetic wave, which results from the transition of the atom or molecule from the initial state of higher energy, E_n, to the final state of lower energy, E_{n+1}. It reads

$$\Delta E = E_n - E_{n+1} = \hbar\nu = \hbar c/\lambda, \qquad (1.2)$$

where $\hbar$ is the universal Planck constant, equal to 6.6256×10^{-3} J-s. Equation (1.2) represents the fundamental quantum relation.

When the emission, that is, the departure of a packet of energy, takes place from one atom, this packet may be subsequently absorbed by another atom, which may be far different from the first one. In other words, that packet arrives in its entirety at a far distant point. This is

1.1. INTRODUCTION

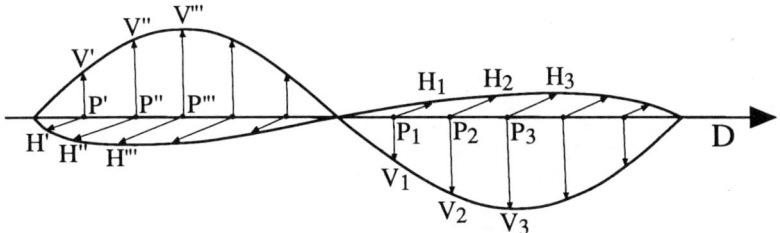

FIG. 1.1. Electromagnetic wave on a wave plane

clearly what would be expected if light consisted of a stream of corpuscles shot out by a luminous surface. One can then state that light in its emission and absorption reveals a corpuscular nature. On the other hand, light in its propagation reveals a wave nature. The two are reconcilable, leading to the so-called "dual characteristics" of light.

1.1.3. THE SIMPLEST WAVE AND CORPUSCLES

If a light source is located at infinity, the wave fronts are plane. Figure 1.1 depicts, at an instant in time, one wavelength interval of the electromagnetic wave on a wave plane, propagating horizontally to the right, as indicated by the arrow D. The wave consists of an identical repetition, both forward and backward, of the single wavelength interval. As time passes, each wave crest, each trough, and each wave contour is propagated to the right with the velocity of electromagnetic waves in vacuum or air. The various lines of propagation, the "rays," are all parallel to D, the particular line of propagation selected for representation. From the corpuscular nature of light, the line D represents a stream of corpuscles advancing in the direction of the arrow, with the same speed.

Let $P', P'', P_1, P_2, \ldots$, be designated equidistant points along the line of propagation. At these points of space, the electric vector has the values $V', V'', \ldots, V_1, V_2, \ldots$, and the magnetic vector has the values $H', H'', \ldots, H_1, H_2, \ldots$. The contour formed by joining the tips of the vectors V is a sine curve, and the same holds true for H. The electric vector is vertical, alternately upward and downward, whereas the magnetic vector is perpendicular to the vertical plane through D and thus horizontal, alternately in and out, from the vertical plane. Any plane perpendicular to D is a wave front; thus, at a given instant in time, V and H possess the same value throughout the plane at P'', the vector V has the same value of V'' pointing upward, and the vector H would have a number of vectors such as H″ pointing perpendicularly out from the page. The vectors V and H are perpendicular to each other and both are, in turn, perpendicular to the direction of propagation. Furthermore, the variation that each vector undergoes is perpendicular to D. The waves belong to the category of being purely transverse. With D always to the right, the directions of D, V, and H, in this cyclic order, form a right-handed system. Noted that the V wave and the H wave jointly constitute the electromagnetic wave. Neither can exist alone.

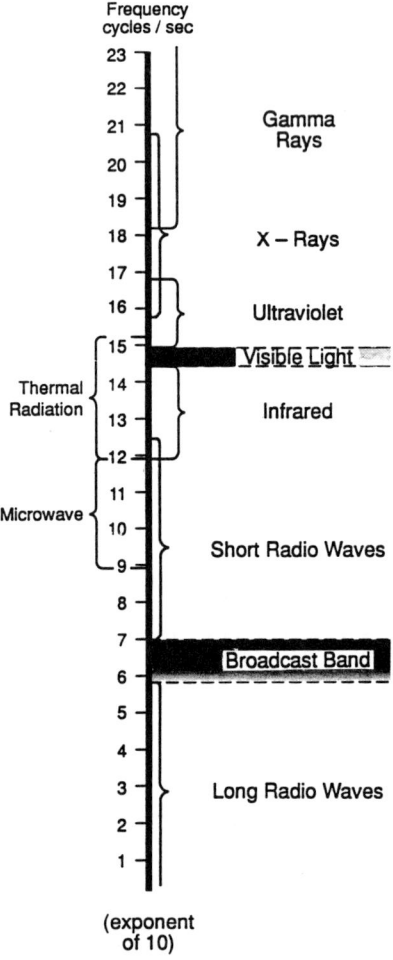

FIG. 1.2. The electromagnetic spectrum

1.2. DEFINITIONS AND LAWS REGARDING THERMAL RADIATION

1.1.4. Electromagnetic Spectrum

The types of electromagnetic radiation can be classified according to their wavelength in vacuum. A chart of the radiation spectrum is shown in Fig. 1.2. The range extends from a wavelength of 3×10^9 cm for the longest electrical waves down to about 3×10^{-15} cm for the shortest wavelengths of cosmic rays, or in frequency from 10 to 10^{25} Hz.

All bodies continuously emit radiation, but heating effects can only be detected if a body's wavelength falls within the spectrum region between 0.1 and 100 μm. The visible range as light is within the narrow band from 0.38 to 0.76 μm.

1.2. Definitions and Laws Regarding Thermal Radiation

1.2.1. Spectroradiometric Curves

Radiation is emitted by bodies by virtue of their temperature. Its importance in thermal calculations is limited to the wavelengths ranging from 0.1 to 100 μm. The total quantity of radiation emitted by a body per unit area and time is called the *total emissive power E*. It depends on the temperature and the surface characteristics of the body. The amount of radiation with certain wavelength λ emitted by a body is referred to as *monochromatic emissive power E_λ*. At any particular temperature, E_λ is different at various wavelengths.

It is now appropriate to introduce an ideal radiator, or blackbody, in radiation. Like the ideal gas, the blackbody is used as a standard with which the radiation characteristics of other bodies are compared. It is a perfect absorber (absorbing all radiation incident upon it) and a perfect emitter, capable of emitting, at any specified temperature, the maximum possible amount of thermal radiation at all wavelengths.

There are three basic laws regarding the emission of radiation from a blackbody:

1. *Planck's law*: In the year 1900, Max Planck derived a relationship showing the spectral distribution of monochromatic emissive power for a blackbody, $E_{b\lambda}$, by means of his quantum theory. It can be expressed as

$$E_{b\lambda} = \frac{2\pi \hbar c^2}{\lambda^5 [\exp(\hbar c / \lambda k T) - 1]}. \qquad (1.3)$$

Here, T denotes the temperature of the blackbody, and k is the universal Boltzmann constant, equal to 1.3805×10^{-3} J/K. Equation

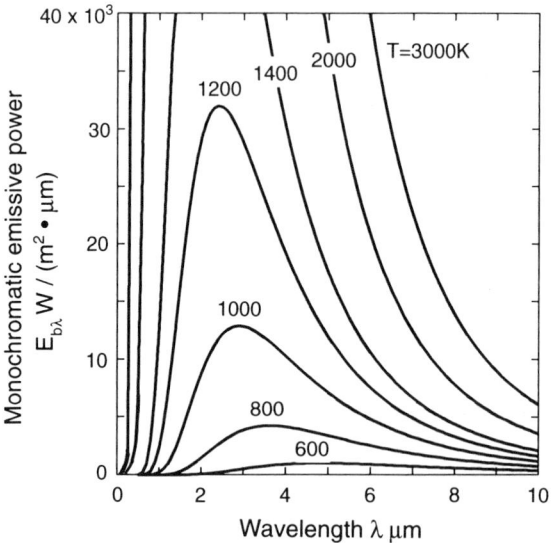

FIG. 1.3. Monochromatic emissive power of a blackbody

(1.3) is graphically depicted in Fig. 1.3 for various temperatures. The curves are called *spectroradiometric curves*.

2. *Wien's displacement law*: The spectroradiometric curve has a peak at the wavelength λ_{max}. Wien's displacement law describes the relationship between λ_{max} and the body temperature T as

$$\lambda_{max} T = 2897.6 \ \mu m \ K. \tag{1.4}$$

The major portion of radiation is emitted within a relatively narrow band to both sides of the peak. For example, the sum, with surface temperature of approximately 6000 K, emits more than 90% of its total radiation between 0.1 and 3 μm, whereas the maximum wavelength is 0.48 μm (with its peak emissive power of $E_{b\lambda_{max}} = 1.03 \times 10^8$ W/m^2).

3. *Stefan-Boltzmann law*: The area under the spectroradiometric curve is the total amount of radiation emitted over all the wavelength, E_b. Mathematically, one writes

$$E_b = \int_0^\infty E_{b\lambda} \, d\lambda. \tag{1.5}$$

Upon the substitution of Eq. (1.3), Eq. (1.5) yields

$$E_b = \sigma T^4. \tag{1.6}$$

1.2. DEFINITIONS AND LAWS REGARDING THERMAL RADIATION

Equation (1.6) expresses a quantitative relationship between the temperature and the total emissive power of a blackbody. It is called the *Stefan-Boltzmann law* and is sometimes referred to as the fourth power law.

1.2.2 RADIATION INTENSITY

Consider radiation emitted uniformly in all directions from an infinitesimal area, dA_1, on surface A_1. The radiation is intercepted by an infinitesimal area dA_2 on a hemispherical surface A_2, which is centered at dA_1 with radius r, as illustrated in Fig. 1.4. We can empirically derive that the rate of radiative heat transfer from dA_1 to dA_2 is proportional to the emitted surface dA_1; the solid angle from dA_1 extending in the direction of dA_2, $d\Omega$; and the cosine of the zenith angle η, $\cos \eta$. That is,

$$dq_{1-2} \propto dA_1 \, d\Omega \cos \eta.$$

The physical observation can be expressed in mathematical term as

$$dq_{1-2} = I \, dA_1 \, d\Omega \cos \eta, \qquad (1.7)$$

in which I is the proportionality constant, named radiation intensity. Equation (1.7) is called *Lambert's cosine law*. For a gray surface, the radiation intensity is a constant, being independent of direction. In general, $I(\theta, \phi) \, d\Omega$ signifies the energy emitted per unit area per unit time into a solid angle $d\Omega$, centered around a direction that can be defined in terms of the zenith angle θ and the azimuthal angle ϕ in the spherical coordinate system.

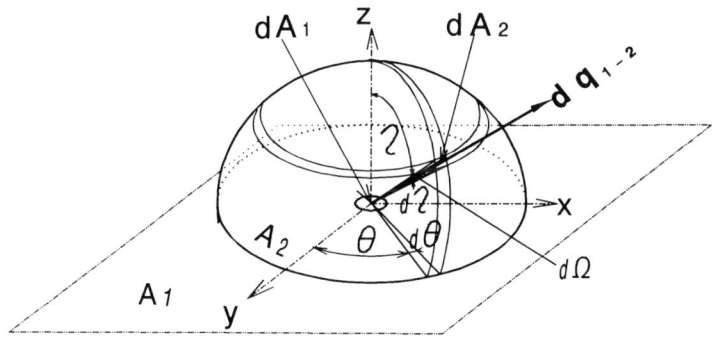

FIG. 1.4. Angles of radiative energy emission from a solid wall

1.2.3. Relationship between Emissive Power and Radiation Intensity

Equation (1.7) divided by dA_1, and defining dq_{1-2}/dA_1 as dE yields

$$dE = I d\Omega \cos \eta. \qquad (1.8)$$

This equation is integrated over the entire hemisphere to give

$$E = \int_\Omega I d\Omega \cos \eta = \int_0^{2\pi} d\theta \int_0^{\pi/2} I \cos \eta \sin \eta \, d\eta. \qquad (1.9)$$

Because I is a constant, we obtain

$$E = \pi I. \qquad (1.10)$$

This relationship also applies to a blackbody and a monochromatic wave:

$$E_b = \pi I_b, \qquad (1.11)$$

$$E_\lambda = \pi I_\lambda. \qquad (1.12)$$

Equations (1.10)–(1.12) relate thermodynamics (LHS) to optics (RHS).

1.2.4. Radiative Characteristics of Solid Surfaces

When a beam of radiation is incident on a surface, a fraction of the incident beam flux, G (W/m^2), is reflected, a fraction is absorbed, and the remaining is transmitted through the solid, as depicted in Fig. 1.5. Let R, A, and T be the fractions of reflection, absorption, and transmission,

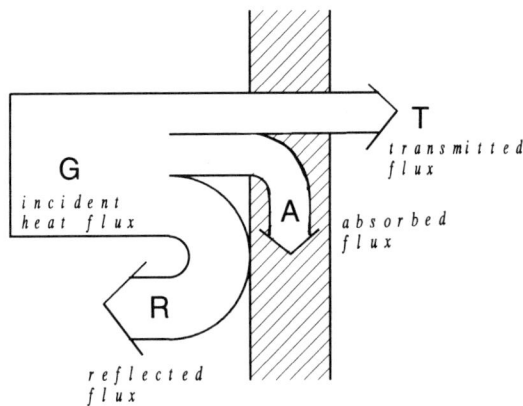

FIG. 1.5. Radiative characteristics of a solid surface

1.2. DEFINITIONS AND LAWS REGARDING THERMAL RADIATION

called *reflective power, absorptive power, and transmissive power*, respectively. The heat balance requires

$$G = R + A + T \quad \text{(all in W/m}^2\text{)}. \tag{1.13}$$

Both sides of the equation are divided by G to give

$$1 = \frac{R}{G} + \frac{A}{G} + \frac{T}{G}.$$

These ratios are defined as

$$\rho = \frac{R}{G}, \quad \alpha = \frac{A}{G}, \quad \tau = \frac{T}{G}, \tag{1.14}$$

where ρ, α, and τ are called the *reflectivity, absorptivity*, and *transmissivity*, respectively. One can write

$$\rho + \alpha + \tau = 1. \tag{1.15}$$

For $\tau = 0$ (in most engineering applications), Eq. (1.15) becomes

$$\rho + \alpha = 1. \tag{1.16}$$

Irrespective of incident radiation, any surface at a temperature above 0 K irradiates energy. The ratio of its emissive power, E, to the emissive power of a blackbody at the same temperature, E_b, is defined as the emissivity ε:

$$\varepsilon = E/E_b \quad (0 \leq \varepsilon \leq 1.0). \tag{1.17}$$

A surface with ε independent of λ is called a *gray surface*. It is a diffuse surface with uniform hemispherical radiation intensity. For radiative energy of a monochromatic wave, λ, the same definition applies:

$$\varepsilon_\lambda = E_\lambda/E_{b\lambda} \quad (0 \leq \varepsilon_\lambda \leq 1), \tag{1.18}$$

where ε_λ is called the monochromatic emissivity. It is a physical property whose magnitude varies with the material and characteristics (such as color, roughness, cleanness, etc.) of the surface. When each variable, ε_λ, E_λ, and $E_{b\lambda}$, is separately integrated with respect to the wavelength from zero to infinity, Eq. (1.18) yields Eq. (1.17). A blackbody is a special case of a gray surface with $\varepsilon = 1.0$.

1.2.5. KIRCHOFF'S LAW AT SOLID SURFACES

Consider n surfaces enclosed by one insulated surface, as shown in Fig. 1.6. Let G be the radiant heat flux from the enclosing surface, which is equivalent to the incident heat flux to all enclosed surfaces. Powers E_1, $E_2, \ldots, E_n$ are, respectively, the emissive powers of the surfaces $1, 2, \ldots, n$,

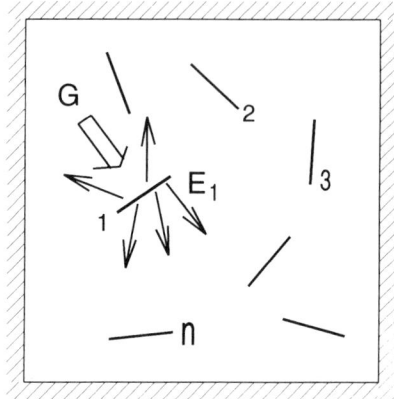

Fig. 1.6. Thermal balance for solid surfaces

which have emissivities of $\varepsilon_1, \varepsilon_2, \ldots, \varepsilon_n$, respectively. Under thermal equilibrium conditions of the system, the enclosing surface and the enclosed surfaces are at the same temperature. The heat balance on surface 1 gives

$$\alpha_1 G = E_1 = \varepsilon_1 E_{b1}. \tag{1.19}$$

The enclosing surface is a blackbody by nature,

$$G = E_b. \tag{1.20}$$

By virtue of thermal equilibrium, $E_b = E_{b1}$. A combination of Eqs. (1.19) and (1.20) yields

$$\alpha_1 = \varepsilon_1. \tag{1.21}$$

Similarly, one obtains, for all other surfaces,

$$\alpha_2 = \varepsilon_2, \ldots, \alpha_n = \varepsilon_n. \tag{1.22}$$

This constitutes the first statement of Kirchhoff's radiation law—that emissivity and absorptivity are equal for a gray surface. The second statement follows—that a blackbody has

$$\alpha = \varepsilon = 1. \tag{1.23}$$

Equation (1.23) describes a blackbody as a perfect absorber as well as a perfect emitter, irrespective of radiative direction or wavelength.

1.3. Definitions and Laws Regarding Gas Radiation

Monatomic gases such as oxygen, nitrogen, and dry air emit very little radiative energy, and can thus be considered transparent. In contrast, diatomic gases such as CO_2, H_2O, CCCO, SO_2, NO, and CH_3, which are contained in combustion gases, do emit and absorb radiative energy in certain long wavelength ranges (i.e., selective radiation). They can be regarded as transparent outside these ranges. Those gases that absorb and emit radiative energy are called thermal radiative gases.

1.3.1. Gray Gases

To determine the propagation of radiative energy in thermal radiative gases, it is customary to divide the radiative energy into one component within the wavelength range of absorption, and the other within the wavelength range of no absorption, which are then separately analyzed. Such a task has proved to be tedious. An engineering approach is to treat it like the case of a gray surface, with neither the absorptive nor emissive characteristics varying with the wavelength. The gas having these characteristics is called *gray gas*.

1.3.2. Absorption Coefficient K (m^{-1})

The attenuation of radiative energy inside a gas of infinitesimally thin thickness dS is proportional to the radiation intensity I and the thickness dS, and can be expressed as

$$dI = -K I dS. \qquad (1.24)$$

Here, the proportionality constant K is called the *gas absorption coefficient*. The equation is valid, exactly, for only a monochromatic ray whose absorptivity is a function of the wavelength, temperature, and pressure. Hence, the energy propagation is treated using the conventional radiative analysis. In the case of radiative energy of a broad wavelength band, Eq. (1.24) is valid only approximately.

Consider a gray gas that satisfies Eq. (1.24). The radiant energy, having an intensity of I_0 within the solid angle $d\Omega$, enters into a gas volume of thickness S and cross-sectional area dF, as illustrated in Fig. 1.7. It is attenuated and exists from the left surface of the gas volume with an intensity of I. The radiation intensity I can be obtained by integrating Eq. (1.24) as

$$I d\Omega dF = I_0 d\Omega dF e^{-KS}.$$

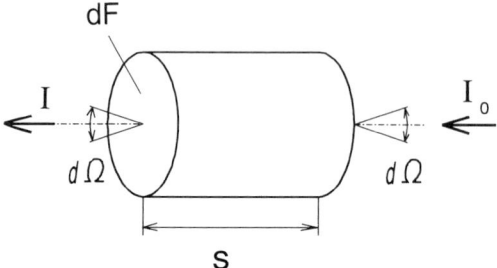

FIG. 1.7. Radiative energy absorption through a gas layer

The LHS of this equation expresses the radiative energy within the solid angle $d\Omega$ that is emitted from the left surface. The RHS signifies the radiative energy within the solid angle $d\Omega$ that exists from the left surface after the radiant energy incident on the right surface is attenuated in the gas volume. The equation leads to Beer's law, which expresses the attenuation of radiant energy inside a gas volume of thickness S as

$$I = I_o e^{-KS}. \tag{1.25}$$

Here, the physical units of K and S are inverse meters and meters, respectively. Their product KS becomes dimensionless, and is named absorptive distance or optical length.

1.3.3. Directional Emissivity $\varepsilon_G(S)$

Consider radiant energy emitted from within a gas volume of thickness S, measured in an arbitrary direction. The gas volume is at a uniform temperature T. Let $I(S)$ be the radiation intensity exiting from a surface of the gas volume in the same direction, and I_b be that emitted from the surface of the gas volume, supposing that it is a solid surface at the same temperature. Then, directional emissivity is defined as the ratio of $I(S)$ to I_b:

$$\varepsilon_G(S) = \frac{I(S)}{I_b} = \frac{\pi I(S)}{E_b} = \frac{\pi I(S)}{\sigma T^4}. \tag{1.26}$$

Hottel [20] presents the values of $\varepsilon_G(S)$ for CO_2 and H_2O in graphical form as functions of S, pressure, and temperature.

1.3. DEFINITIONS AND LAWS REGARDING GAS RADIATION

1.3.4. DIRECTIONAL ABSORPTIVITY $\alpha_G(S)$

Consider the radiant energy of intensity I_0 entering an isothermal gas volume of thickness S, measured in an arbitrary direction. It attenuates by $(I_0 - I)$ after being absorbed within the gas volume. The directional absorptivity $\alpha_G(S)$ is defined as

$$\alpha_G(S) = \frac{I_0 - I}{I_0}. \tag{1.27}$$

Substituting Eq. (1.25), the expression is reduced to

$$\alpha_G(s) = 1 - e^{-KS}. \tag{1.28}$$

For a gray gas whose absorption coefficient K is independent of the wavelength, Eqs. (1.25), (1.26), and (1.28) combine to give

$$\varepsilon_G(S) = \alpha_G(S) = 1 - e^{-KS}. \tag{1.29}$$

With the value of $\varepsilon_G(S)$ evaluated graphically from Hottel's chart, Eq. (1.29) can be used to determine the absorption coefficient of a gray gas K with a volume of thickness S.

1.3.5. TOTAL RADIANT ENERGY IRRADIATED FROM AN ISOTHERMAL GAS VOLUME

Now, consider the total radiant energy irradiated from an isothermal gas volume in all directions (solid angle $= 4\pi$). The radiant energy emitted from an infinitesimal volume of $dV = dS\,dF$ within an infinitesimal solid angle $d\Omega$, as depicted in Fig. 1.8

$$jF\,dS\,d\Omega = jdV d\Omega. \tag{1.30}$$

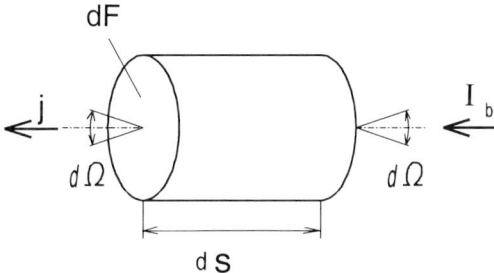

FIG. 1.8. Radiative energy emission from a gas volume

Here, j is the radiant energy irradiated from the surface of a unit volume of the gas volume, per unit time per unit solid angle. Since an infinitesimally small gas volume ($dV \to 0$) is considered, it is postulated that self-absorption does not take place before a portion of emitted radiant energy exists from the gas volume. From Kirchhoff's equation, j is related to the blackbody radiation intensity incident on the gas volume I_b by

$$j = KI_b. \qquad (1.31)$$

This equation can be proved by considering the heat balance of an infinitesimally small volume of a gas block confined in a cavity surrounded by a solid wall at temperature T. Being a cavity, the radiation intensity in it is equal to that of a blackbody at temperature T. Under the thermal equilibrium condition, the radiant energy being absorbed by the small gas volume is equal to that being emitted. Equation (1.30) says the energy irradiated from an infinitesimal volume $dV = dFdS$ into an infinitesimal solid angle $d\Omega$ is $jdFdSd\Omega$. Equation (1.24) yields the energy entering into the infinitesimal volume through an entrance cross-sectional area of dF at a solid angle of $d\Omega$, $I_b dFd\Omega$, and being absorbed inside the infinitesimal volume with a thickness of dS, to be $(I_b dFd\Omega)KdS$. By equating the energy irradiated to the energy absorbed, Eq. (1.31), representing Kirchhoff's radiation law, is established.

Combining Eqs. (1.11) and (1.9), the energy irradiated from a gas volume having an infinitesimal volume $dV = dFdS$ through an infinitesimal solid angle $d\Omega$ is found to be

$$jdVd\Omega = KI_b dVd\Omega$$

$$= K(E_b/\pi)dVd\Omega$$

$$= K(\sigma T^4/\pi)dVd\Omega.$$

Hence, total energy e, which is irradiated from the gas volume of an infinitesimal volume dV in all directions (solid angle $= 4\pi$), can be expressed as

$$e = 4\pi K(\sigma T^4/\pi)dV = 4K\sigma T^4 dV. \qquad (1.32)$$

Note that this equation is derived for a gas volume with such a small volume that the self-absorption is negligible. If the gas volume is of finite size with self-absorption, the energy irradiated from the gas volume is diminished accordingly, as is illustrated later by Eq. (3.3) in Chapter 3.

1.4. Definitions and Laws Regarding Radiation from Gas–Particle Mixtures

In ordinary engineering problems, the scattering of radiation in a gas can be ignored. However, if many solid particles or liquid droplets (to be called particles as a general term) are suspended in the gas, they may cause an appreciable scattering of radiation.

1.4.1. Attenuation Constant a (m^{-1})

The attenuation of radiation energy in a particle-containing gas is induced by absorption by the gas and particles and also by scattering due to the particles. It is described by an equation that is similar in form to Eq. (1.25) representing Beer's law for radiant energy attenuation in the absence of scattering:

$$I = I_0 e^{-aS}, \tag{1.33}$$

where a is the attenuation constant for the gas–particle mixture, and aS is the attenuation distance. The attenuation constant a is the sum of the gas absorption coefficient K, particle absorption cross section σ_a, and scattering cross section of particles σ_s:

$$a = K(1 - N\pi d^3/6) + \sigma_a + \sigma_s. \tag{1.34}$$

For the cases of particle diameters being much larger than the wavelength, and packed layers with a low volume density of particles, the absorption and scattering cross sections of particles in a common particle-containing gas can be expressed as

$$\sigma_a = N\varepsilon_p \pi d^2/4 \tag{1.35}$$

and

$$\sigma_s = N(1 - \varepsilon_p)\pi d^2/4. \tag{1.36}$$

For smaller particles, σ_a and σ_s can be determined by evaluating the absorptive efficiency factor or the scattering efficiency factor by means of the Mie theory.

1.4.2. Scattering Albedo ω

The scattering albedo is defined as the fraction of scattering in the total attenuation of radiant energy that passes through a particle-containing gas.

18   THERMAL RADIATION

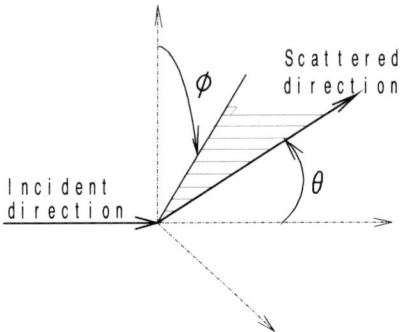

FIG. 1.9. Definition of scattering angles

It reads

$$\omega = \frac{\sigma_s}{a} = \frac{\sigma_s}{K(1 - N\pi d^3/6) + \sigma_a + \sigma_s}. \qquad (1.37)$$

1.4.3. SCATTERING PHASE FUNCTION $\Phi(\theta, \phi)$

Figure 1.9 shows radiant energy is scattered in the (θ, ϕ) direction with respect to the incident direction. The relative intensity distribution of the scattered radiant energy is expressed by the ratio of the scattered radiation intensity to the intensity of isotropic scattering in the same direction. The value of Φ is unity in the isotropic scattering case. In general, the averaged value of Φ over the entire spherical angle of 4π is equal to unity. That is,

$$\frac{1}{4\pi} \int_{\Omega_s = 4\pi} \Phi(\theta, \phi) d\Omega_s = 1. \qquad (1.38)$$

The scattering phase function is purely a function of θ for a particle, such as a sphere, where scattering characteristics are independent of the circumferential direction ϕ.

The scattering phase function of a gray sphere with a radius sufficiently greater than the wavelength can be expressed as

$$\phi(\theta) = \frac{8}{3\pi}(\sin\theta - \theta\cos\theta). \qquad (1.39)$$

Mie's, or Rayleigh's, scattering theory is applied to determine the scattering phase function of particles with radius similar to, or smaller than, the wavelength, respectively.

Chapter 2

Radiation Heat Transfer

With the introduction of thermal radiation, which explains the radiative characteristics of a single surface, gas, and particle, we now direct our attention toward radiative heat exchange.

2.1. Basic Equations

This section presents the basic equations employed in the radiative analyses of heat transport by radiation, convection, and conduction in media having radiation, absorption and scattering characteristics. Problems often encountered in the solution process are also described. For simplicity, gas and particles are considered to be at the same temperature.

2.1.1. Energy Equation

The energy balance equation for an infinitesimal element reads

$$\rho C_p \frac{DT}{D\tau} = \frac{Dp}{D\tau} + \nabla \cdot (k\nabla T - \mathbf{q}_r) + q_h + \Phi, \qquad (2.1)$$

where $\rightarrow$ indicates a directional quantity; ρ, density; C_p, specific heat under constant pressure; T, temperature; τ, time; p, pressure; k, thermal conductivity; $\mathbf{q}_r$, radiant heat flux vector; q_h, chemical heat generation rate; and Φ, viscous heat dissipation rate. In addition, $D/D\tau$ is the total derivative defined as

$$\frac{D}{D\tau} = \frac{\partial}{\partial t} + \mathbf{v} \cdot \nabla \qquad (2.2)$$

The symbol ∇, called del, is defined as

$$\nabla = \mathbf{i}\frac{\partial}{\partial x} + \mathbf{j}\frac{\partial}{\partial y} + \mathbf{k}\frac{\partial}{\partial z}$$

where **i**, **j**, and **k** are the unit vectors in the x, y, and z directions, respectively. The LHS of Eq. (2.1) represents the local acceleration and convective terms, and the RHS consists of the pressure work, conduction, radiation, chemical heat generation, and viscous heat dissipation, respectively. The third term is called the *radiative heat flux vector*, which represents the total radiant heat energy transferred across a certain cross section. It can be expressed as

$$\mathbf{q}_r = \mathbf{i} q_{rx} + \mathbf{j} q_{ry} + \mathbf{k} q_{rz}$$

$$= \mathbf{i} \int_0^{4\pi} I' \cos \alpha \, \partial \Omega + \mathbf{j} \int_0^{4\pi} I' \cos \beta \, \partial \Omega + \mathbf{k} \int_0^{4\pi} I' \cos \gamma \, \partial \Omega, \quad (2.3)$$

where q_{rx}, q_{ry}, q_{rz} are the net radiative heat flux and **i, j, k** are the unit vectors in the x, y, z directions, respectively, and α, β, γ are the angles between the directions of the I' and the x, y, z axes, respectively.

2.1.2. Divergence of Radiative Heat Flux Equation

The $\nabla \cdot \mathbf{q}'_r$ term in Eq. (2.1) signifies the divergence of radiative heat flux, that is, the net radiant energy emitted from a unit volume. It is a scalar quantity (i.e., no directional characteristics), a function of position, and can be expressed as

$$\nabla \cdot \mathbf{q}'_r = 4\pi \int_0^\infty \bigg(K_\lambda(\lambda, T) I'_{\lambda b}(\lambda)$$

$$- \{[K_\lambda(\lambda, T) + \sigma_{a\lambda}(\lambda)] + \sigma_{s\lambda}(\lambda)\} \bar{I}_{\lambda b}(\lambda)$$

$$+ \frac{\sigma_{s\lambda}(\lambda)}{4\pi} \int_0^{4\pi} \int_0^{4\pi} I'_\lambda(\lambda, \Omega_i) \Phi(\lambda, \Omega, \Omega_i) \, d\Omega_i \, d\Omega \bigg) d\lambda. \quad (2.4)$$

The terms of the integrand on the RHS represent the self-radiation, the attenuation of the incident radiation by absorption and scattering (inside the brackets), and the emission resulting from scattering (the double integral), respectively. The situation is graphically illustrated in Fig. 2.1. $\bar{I}_{\lambda b}(\lambda)$ is obtained from

$$\bar{I}_{\lambda b}(\lambda) = \frac{1}{4\pi} \int_0^{4\pi} I'_\lambda(\lambda) \, d\Omega, \quad (2.5)$$

which represents the average value of all radiant energy incident on a point.

2.1. BASIC EQUATIONS

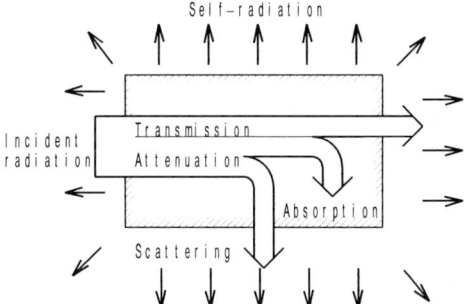

FIG. 2.1. Radiative energy balance in a gas element

2.1.3. TRANSPORT EQUATIONS

The intensity of radiant energy incident on a point on the abscissa, at optical length of κ_λ and in the Ω direction, can be expressed in the integral form as

$$I'_\lambda(\kappa_\lambda, \Omega) = I'_\lambda(0) \exp(-\kappa_\lambda)$$
$$+ \int_0^{\kappa_\lambda} i'_\lambda(\kappa^*_\lambda, \Omega) \exp[-(\kappa_\lambda - \kappa^*_\lambda)] \, d\kappa^*_\lambda. \quad (2.6)$$

This is the transport equation in integral form. The first term on the RHS is the portion of the incident radiation from the boundary that is not attenuated but instead reaches the gas element. The second term represents the portion of radiant energy emitted by all other gas elements (i.e., expressed by the source term, i'_λ) that reaches the element. The transport equation can also be expressed in the differential form as

$$\frac{dI'_l(\kappa_\lambda, \Omega)}{d\kappa_\lambda} + I'_\lambda(\kappa_\lambda) = i'_\lambda(\kappa_\lambda, \Omega). \quad (2.7)$$

Here, the optical length κ_λ is defined as

$$\kappa_\lambda(s) = \int_0^s a_\lambda(s^*) \, ds^* = \int_0^s [K_\lambda(s^*) + \sigma_a(s^*) + \sigma_s(s^*)] \, ds^* \quad (2.8)$$

2.1.4. DEFINITION OF SOURCE FUNCTION OF RADIATION INTENSITY

The source function $i_\lambda{}'(\kappa_\lambda, \Omega)$ is defined as an increase in the radiation intensity per unit thickness in the Ω direction at the location on the abscissa at an optical length of κ_λ. It is, similar to the radiation intensity, a function of the location, direction, and wavelength, and can be expressed as

$$i'_\lambda(\kappa_\lambda, \Omega) = (1 - \omega_\lambda) I'_{\lambda b}(\kappa_\lambda)$$
$$+ \frac{\omega_\lambda}{4\pi} \int_0^{4\pi} I'_\lambda(\kappa_\lambda, \Omega_i) \Phi(\lambda, \Omega, \Omega_i) \, d\Omega_i. \quad (2.9)$$

The first term on the RHS denotes the local self-radiation at κ_λ, and the second term is the directional component of a change in the radiant energy in the Ω direction due to scattering.

The conventional radiation heat transfer analysis calls for a direct solution of the basic heat balance equation in an integrodifferential form. The independent variables at each point in the system include:

1. The temperature T, which determines the self-emission from each point into the surrounding area
2. The source function i'_λ, which includes the self-emission and the scattering from each point into the surrounding area
3. The radiation intensity I'_λ, defined as the radiant energy passing through each point in each direction
4. The radiant heat flux $\mathbf{q}_r$, which is the total radiant energy passing through a cross section integrated in all directions around the point.

Among the four variables, the source function and the radiation intensity are related via (i.e., coupled by) Eqs. (2.6) and (2.9). The radiant heat flux is determined from Eq. (2.4) using the two variables and is related to convection and conduction via Eq. (2.1). Therefore, to obtain the solution for a heat transfer problem with a combination of conduction, convection, and radiation mechanisms, the four coupled equations [Eqs. (2.1), (2.4), (2.6), and (2.9)] must be simultaneously solved. In principle, the problem can be solved, because it has four variables with four equations. However, because radiant energy is transferred by electromagnetic waves, the source function i'_λ and radiation intensity I'_λ at each point are expressed as the integrations of the effects at all other points in the system. They are functions of not only the location but also the direction. As a result, the analysis becomes extremely difficult. It is a general practice, in solving the equations analytically or numerically, to impose certain assumptions to simplify the problem. The most common simplifications include these:

1. Scattering is assumed to be isotropic, in order to avoid the difficulty due to the directional dependence of these variables
2. Reduction in the number of dimensions, for example, to one dimension, as in the two-flux method (in the following section)

2.2. EXISTING METHODS OF SOLUTIONS

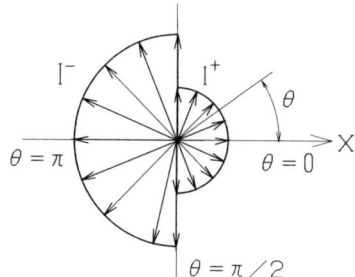

FIG. 2.2. Schuster-Schwartzschild approximation of intensities

3. Constant physical properties
4. Severance of coupling of heat transfer modes by neglecting the effects of convection and conduction.

2.2. Existing Methods of Solutions

Both the two-flux and zone methods have been used extensively to solve the transport equations. This section presents simple explanations of their principles and special features. Examples are provided to demonstrate the analysis of a one-dimensional system using each method.

2.2.1. Two-Flux Method

Consider a one-dimensional system in the x axis, as depicted in Fig. 2.2. Let I^+ and I^- be the radiation intensities in the positive and negative directions of the coordinate axis, respectively. The two-flux method approximates that the angular distributions of both I^+ and I^- are uniform and constant (isotropic), but with different values. The assumption of one dimensionality makes it possible to solve the complete equation of transfer with relative simplicity. Two solution methods have been popularly employed: the Schuster-Schwarzschild approximation and the Milne-Eddington approximation [21]. In dealing with the radiant energy that travels through a gas layer of infinitesimal thickness dx, perpendicular to the coordinate axis, both assume that for one-dimensional energy transfer, the intensity in the positive direction is isotropic and in the negative direction is also isotropic at a different value, as seen in Fig. 2.2. By multiplying the radiation intensity by $\cos\theta$ followed by integrating the resulting expression over the entire solid angle, one obtains the radiation heat flux, as defined in Eq. (2.3). The Milne-Eddington approximation is used to compute heat fluxes.

This section explains the Schuster-Schwarzschild approximation, the simpler of the two methods. The equation of transfer is written for the intensity in each hemisphere as

$$\frac{dI_\lambda^+(x)}{d(x/\cos\theta)} = -K_\lambda I_\lambda^+(x) + K_\lambda I'_{\lambda b}(x) \qquad (0 \leq \theta \leq \pi/2), \quad (2.10)$$

$$\frac{dI_\lambda^-(x)}{d(x/\cos\theta)} = -K_\lambda I_\lambda^-(x) + K_\lambda I'_{\lambda b}(x) \qquad (\pi/2 \leq \theta \leq \pi). \quad (2.11)$$

These are derived by substituting the source function, which is obtained from Eq. (2.9) with $\omega_\lambda = 0$ (for no scattering), into Eq. (2.7). From the isotropic postulation, I_λ^+ and I_λ^- do not depend on θ. These equations are now integrated over their respective hemispheres to yield

$$-\frac{1}{2K_\lambda} \frac{dI_\lambda^+(x)}{dx} = I_\lambda^+(x) - I'_{\lambda b}(x), \quad (2.12)$$

$$-\frac{1}{2K_\lambda} \frac{dI_\lambda^-(x)}{dx} = I_\lambda^-(x) - I'_{\lambda b}(x). \quad (2.13)$$

These equations, with appropriate boundary conditions, are solved to determine the distributions of temperature and heat flux.

As an example, let us consider radiation heat transfer between two parallel, infinite, gray plates at different temperatures, as illustrated in Fig. 2.3. The space between the plates contains a nonheat-generating, nonscattering, gray gas. In Eqs. (2.12) and (2.13), the distance x is replaced by an optical thickness defined as $\kappa = K_\lambda x$; the subscript λ in all terms is deleted with the assumption of a gray gas, and the equations are integrated from both plates to the location κ inside the gas space, yielding the equation of transfer in the integral form for the two-flux method:

$$I^+(\kappa) = I^+(0)\exp(-2\kappa) + 2\int_0^\kappa I'_b(\kappa^*)\exp[2(\kappa^* - \kappa)]\,d\kappa^*, \quad (2.14)$$

$$I^-(\kappa) = I^-(\kappa_D)\exp[2(\kappa - \kappa_D)] + 2\int_\kappa^{\kappa_D} I'_b(\kappa^*)\exp[2(\kappa - \kappa^*)]\,d\kappa^*. \quad (2.15)$$

The expressions are similar to Eq. (2.6).

Next we relate the gas temperature T to I^+ and I^- at each location κ, for the purpose of determining the temperature distribution. Since there is no internal heat generation within the gas, the emission and absorption of radiant energy must be balanced in any gas block. This yields

$$4K\sigma T^4(\kappa) = K\int_0^{4\pi} I'(\kappa)\,d\Omega = K2\pi[I^+(\kappa) + I^-(\kappa)] \quad (2.16)$$

2.2. EXISTING METHODS OF SOLUTIONS

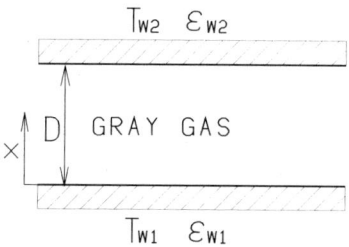

FIG. 2.3. Analytical model

or

$$\sigma T^4(\kappa) = \frac{\pi}{2}[I^+(\kappa) + I^-(\kappa)]. \tag{2.17}$$

Equation (2.3) gives the radiant heat flux vector as

$$q_r = \int_0^{4\pi} I' \cos\theta \, d\Omega = \int_0^{\pi} I' \cos\theta \, (2\pi \sin\theta) \, d\theta$$

$$= 2\pi \left[I^+(\kappa) \int_0^{\pi/2} \cos\theta \sin\theta \, d\theta + I^-(\kappa) \int_{\pi/2}^{\pi} \cos\theta \sin\theta \, d\theta \right]$$

$$= \pi [I^+(\kappa) - I^-(\kappa)]. \tag{2.18}$$

In the absence of internal heat generation in the gas, q_r is a constant irrespective of κ. Hence, one can write

$$q_r(\kappa) = q(0) = \pi[I^+(0) - I^-(0)]. \tag{2.19}$$

When Eqs. (2.14) and (2.15) are substituted into Eqs. (2.17) and (2.18) followed by the application of $I'_b(\kappa) = \sigma T^4(\kappa)/\pi$, one gets both the heat balance equation and the heat flux equation as a function of only temperature:

$$\sigma T^4(\kappa) = \frac{1}{2}\left\{ \pi I^+(0) \exp(-2\kappa) + 2\int_0^{\kappa} \sigma T^4(\kappa^*) \exp[2(\kappa^* - \kappa)] \, d\kappa^* \right.$$

$$\left. + \pi I^-(\kappa_D) \exp[2(\kappa - \kappa_D)] + 2\int_{\kappa}^{\kappa_D} \sigma T^4(\kappa^*) \exp[2(\kappa - \kappa^*)] \, d\kappa^* \right\}, \tag{2.20}$$

$$q_r = \pi I^+(0) - \pi I^-(\kappa_D) \exp(-2\kappa_D)$$

$$- 2\int_0^{\kappa_D} \sigma T^4(\kappa^*) \exp(-2\kappa^*) \, d\kappa^*. \tag{2.21}$$

Hence, the temperature distribution $T(\kappa)$ and the heat flux q_r can be obtained upon the specification of the boundary conditions $I^+(0)$ and $I^-(K_D)$, respectively at the upper and lower boundaries of the gas layer.

The boundary conditions $I^+(0)$ and $I^-(\kappa_D)$ can be determined as follows. Because the heat flux from the $\kappa = 0$ surface in the direction of positive κ, $(q_1)^+$, is equal to the sum of the self-radiation of the plate surface and the reflected portion of the heat flux incident on the plate, then:

$$(q_1)^+ = \varepsilon_1 \sigma T_{w1}^4 + (1 - \varepsilon_1)(q_1)^-. \tag{2.22}$$

Because I^+ and I^- are uniform in their respective hemispheres, the following equations are given

$$(q_1)^+ = \pi I^+(0), \tag{2.23}$$

$$(q_1)^- = \pi I^-(0). \tag{2.24}$$

Equations (2.23) and (2.24) are substituted into Eq. (2.22) to yield

$$\pi I^+(0) = \varepsilon_1 \sigma T_{w1}^4 + (1 - \varepsilon_1) \pi I^-(0) \tag{2.25}$$

By setting $\kappa = 0$ in Eq. (2.15), one obtains

$$I^-(0) = I^-(\kappa_D) \exp(-2\kappa_D)$$

$$+ 2\int_0^{\kappa_D} I_b'(\kappa^*) \exp(-2\kappa_D^*) \, d\kappa_D^*. \tag{2.26}$$

Now, Eq. (2.26) is substituted into Eq. (2.25) to get

$$I^+(0) = \frac{\varepsilon_1 \sigma T_{w1}^4}{\pi} + (1 - \varepsilon_1)\left[I^-(\kappa_D) \exp(-2\kappa_D)\right.$$

$$\left. + 2\int_0^{\kappa_D} \frac{\sigma T^4(\kappa^*)}{\pi} \exp(-2\kappa^*) \, d\kappa^*\right]. \tag{2.27}$$

In a similar manner, $I^-(\kappa_D)$ is obtained as

$$I^-(\kappa_D) = \frac{\varepsilon_2 \sigma T_{w2}^4}{\pi} + (1 - \varepsilon_2)\left[I^+(0) \exp(-2\kappa_D)\right]$$

$$+ 2\int_0^{\kappa_D} \frac{\sigma T^4(\kappa^*)}{\pi} \exp[-2(\kappa_D - \kappa^*)\, d\kappa^*]. \tag{2.28}$$

2.2. EXISTING METHODS OF SOLUTIONS

Define the parameters A and B as

$$A = 2\int_0^{\kappa_D} \frac{\sigma T^4(\kappa^*)}{\pi} \exp(-2\kappa^*)\, d\kappa^*, \tag{2.29}$$

$$B = 2\int_0^{\kappa_D} \frac{\sigma T^4(\kappa^*)}{\pi} \exp[-2(\kappa_D - \kappa^*)]\, d\kappa^*. \tag{2.30}$$

Equations (2.27) and (2.28) are then solved to yield

$$I^+(0) = \frac{\dfrac{\varepsilon_1 \sigma T_{w1}^4}{\pi} + (1-\varepsilon_1)A + (1-\varepsilon_1)\exp(-2\kappa_D)\left[\dfrac{\varepsilon_2 \sigma T_{w2}^4}{\pi} + (1-\varepsilon_2)B\right]}{1 - (1-\varepsilon_1)(1-\varepsilon_2)\exp^2(-2\kappa_D)} \tag{2.31}$$

and

$$I^-(\kappa_D) = \frac{(1-\varepsilon_2)\exp(-2\kappa_D)\left[\dfrac{\varepsilon_1 \sigma T_{w1}^4}{\pi} + (1-\varepsilon_1)A\right] + \dfrac{\varepsilon_2 \sigma T_{w2}^4}{\pi} + (1-\varepsilon_2)B}{1 - (1-\varepsilon_1)(1-\varepsilon_2)\exp^2(-2\kappa_D)}. \tag{2.32}$$

Figure 2.4 lists the program TFM, which determines the temperature distribution and the heat flux in the system of Fig. 2.3 by means of the Schuster-Schwarzschild approximation of the two-fluid method. The inputs TW1 and TW2 are, respectively, the wall temperatures T_{w1} and T_{w2} (with the unit of K); ANK, optical thickness of the gas layer κ; EM1 and EM2, the wall emissivities ε_1 and ε_2, respectively; and N, number of element divisions of the gas in the x direction.

An example of the outputs from the program is illustrated in Fig. 2.5, corresponding to the input conditions of $T_{w1} = 1000$ K, $T_{w2} = 500$ K, AKD = 2, EM1 = EM2 = 0.5, and $N = 20$. The last output QND is the dimensionless heat flux defined as

$$\psi = \frac{q_r}{\sigma(T_{w1}^4 - T_{w2}^4)}. \tag{2.33}$$

This is the actual heat flux divided by the radiant heat flux in the gas between the two plates in the absence of energy absorption, $\sigma(T_{w1}^4 - T_{w2}^4)$.

Figure 2.6 compares the exact solution [22] with the results obtained by the Monte Carlo method and the two-flux method for the same inlet conditions as those of Fig. 2.5, except the optical thickness and the wall emissivities $\varepsilon(\varepsilon_1 = \varepsilon_2)$. The Monte Carlo method, which is introduced in Chapter 3, uses the program for radiation heat transfer analysis,

```
1    *******************************************************************************
2    *                                                                             *
3    *                                    TFM                                      *
4    *         1-D ANALYSIS ON RAIDATION HEAT TRANSFER BY TWO-FLUX MODEL           *
5    *                  (THE SCHUSTER-SCHWARZSHILD APPROXIMATION)                  *
6    *******************************************************************************
7          DIMENSION AK(100),TG(100)
8          open ( 6,file='PRN' )
9          write(*,100)
10   100   format(1h ,'input TW1 (K), TW2 (K), AKD, EM1, EM2, N'/)
11         READ(*,*) TW1,TW2,AKD,EM1,EM2,N
12         DAK=AKD/FLOAT(N)
13         DO 1000 I=1,N
14           AK(I)=(FLOAT(I-1)+0.5)*DAK
15   1000  CONTINUE
16         TAV=(TW1+TW2)*0.5
17         DO 1010 I=1,N
18           TG(I)=TAV
19   1010  CONTINUE
20         SBC=5.6687E-8
21         PAI=3.14159
22         IF(EM1.EQ.1.0) THEN
23           AIP0=SBC*TW1**4/PAI
24         END IF
25         IF(EM2.EQ.1.0) THEN
26           AINKD=SBC*TW2**4/PAI
27         END IF
28   5000  CONTINUE
29         IF((EM1.LT.1.0).OR.(EM2.LT.1.0)) THEN
30           A=0.0
31           B=0.0
32           DO 1020 I=1,N
33             A=A+TG(I)**4*EXP(-2.0*AK(I))*DAK
34             B=B+TG(I)**4*EXP(-2.0*(AKD-AK(I)))*DAK
35   1020    CONTINUE
36           A=A*2.0*SBC/PAI
37           B=B*2.0*SBC/PAI
38           IF(EM1.LT.1.0) THEN
39             AIP0=(EM1*SBC*TW1**4/PAI+(1.0-EM1)*A+(1.0-EM1)
40        1         *EXP(-2.0*AKD)*(EM2*SBC*TW2**4/PAI+(1.0-EM2)*B))
41        2         /(1.0-(1.0-EM1)*(1.0-EM2)*EXP(-2.0*AKD)**2)
42           END IF
43           IF(EM2.LT.1.0) THEN
44             AINKD=((1.0-EM2)*EXP(-2.0*AKD)*(EM1*SBC*TW1**4/PAI
45        1         +(1.0-EM1)*A)+EM2*SBC*TW2**4/PAI+(1.0-EM2)*B)
46        2         /(1.0-(1.0-EM1)*(1.0-EM2)*EXP(-2.0*AKD)**2)
47           END IF
48         END IF
49         EPS=-1.0
50         DO 1030 I=1,N
51           TP=TG(I)
52           TI=0.0
53           DO 1040 J=1,N
54             TI=TI+TG(J)**4*EXP(-2.0*ABS(AK(J)-AK(I)))*DAK
55   1040    CONTINUE
56           TG(I)=(TI+PAI*(AIP0*EXP(-2.0*AK(I))+AINKD*EXP(-2.0*
57        1         (AKD-AK(I))))/(2.0*SBC))**0.25
58           EPSI=ABS(TG(I)-TP)/TG(I)
59           IF(EPSI.GT.EPS) THEN
```

FIG. 2.4. Program of two-flux model

RADIAN, listed in Fig. 6.2 on Chapter 6. The element division used in the RADIAN program is the one-dimensional system shown in Fig. 2.7. The upper and lower walls are gray with specified temperatures, whereas the side walls are perfectly reflective walls with zero emissivity. Note that

2.2. EXISTING METHODS OF SOLUTIONS

```
60              EPS=EPSI
61           END IF
62    1030 CONTINUE
63           write(*,*) eps
64           IF(EPS.GE.1.0E-5) GOTO 5000
65           QW=0.0
66           DO 1050 I=1,N
67              QW=QW+TG(I)**4*EXP(-2.0*AK(I))*DAK
68    1050 CONTINUE
69           QW=PAI*(AIPO-AINKD*EXP(-2.0*AKD))-2.0*SBC*QW
70           QND=QW/(SBC*(TW1**4-TW2**4))
71           WRITE(6,200)
72     200 FORMAT(1H ,'1-D ANALYSIS ON RADIATION HEAT TRANFER BY TWO-FLUX MOD
73          1EL')
74           WRITE(6,300) TW1,TW2,AKD,EM1,EM2,N
75     300 FORMAT(1H ,'TW1=',E12.5,'(K)',2X,'TW2=',E12.5,'(K)'/
76          1 ' OPTICAL THICKNESS=',E12.5,2X,'EM1=',E12.5,2X,'EM2=',
77          2 E12.5/' NUMBER OF ELEMENTS=',I3/)
78           DO 1060 I=1,N
79              WRITE(6,400) I,TG(I)
80     400    FORMAT(1H ,3X,'I=',I3,2X,'T=',E12.5,'(K)')
81    1060 CONTINUE
82           WRITE(6,500) QW,QND
83     500 FORMAT(1H ,'QW=',E12.5,'(W/m2)',4X,'QND=',E12.5)
84           STOP
85           END
```

FIG. 2.4. (*Continued*)

the RADIAN program (Fig. 3.31) employed for this analysis was modified to treat the perfectly reflective side walls using the method described in Section 3.3.4.

```
                              TFM
             1-D ANALYSIS ON RADIATION HEAT
                TRANSFER BY TWO-FLUX MODEL
                   I=  1    T=    .92255E+03(K)
                   I=  2    T=    .91672E+03(K)
                   I=  3    T=    .91075E+03(K)
                   I=  4    T=    .90462E+03(K)
                   I=  5    T=    .89834E+03(K)
                   I=  6    T=    .89189E+03(K)
                   I=  7    T=    .88527E+03(K)
                   I=  8    T=    .87846E+03(K)
                   I=  9    T=    .87146E+03(K)
                   I= 10    T=    .86426E+03(K)
                   I= 11    T=    .85685E+03(K)
                   I= 12    T=    .84921E+03(K)
                   I= 13    T=    .84133E+03(K)
                   I= 14    T=    .83319E+03(K)
                   I= 15    T=    .82478E+03(K)
                   I= 16    T=    .81608E+03(K)
                   I= 17    T=    .80705E+03(K)
                   I= 18    T=    .79769E+03(K)
                   I= 19    T=    .78796E+03(K)
                   I= 20    T=    .77783E+03(K)
             QW=   .10212E+05(W/m2)     QND=    .19215E+00
```

FIG. 2.5. Output of TFM program

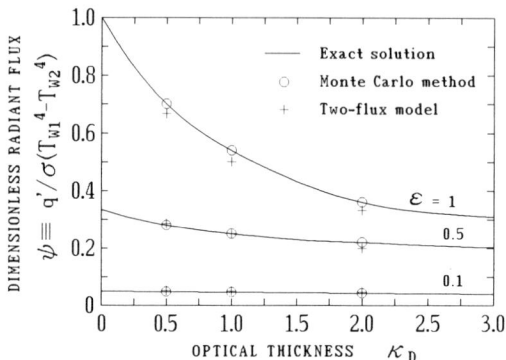

FIG. 2.6. Comparison of results obtained by two-flux model with other methods

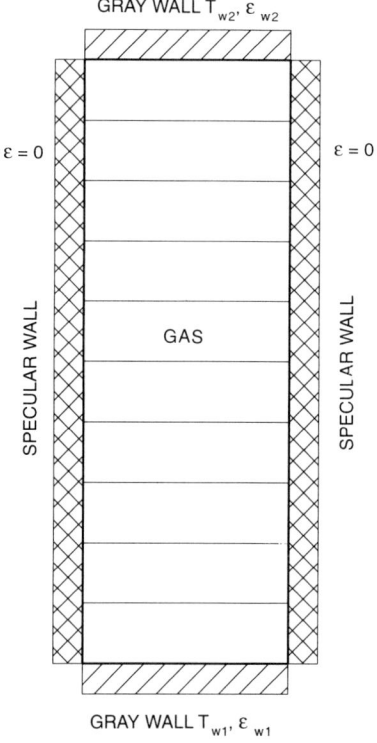

FIG. 2.7. Analytical model

Figure 2.6 shows that the results obtained by the Monte Carlo method agree well with the exact solution. The results by the two-flux method (Schuster-Schwarzschild approximation) also give good agreement with the exact solution, but slightly underpredict in the case of higher wall emissivities and thicker optical thickness.

For one-dimensional systems, the two-flux method can obtain the solution with relative case. However, the method is known to deviate from the exact solutions in the case of a two- or three-dimensional system, or in the presence of scattering, especially strong nonisotropic scattering.

2.2.2. ZONE METHOD

To conduct radiative heat transfer analyses by the zone method, a system is divided into many gas and wall elements, and the temperature is assumed to be constant within each element. To obtain the radiative energy exchange between the elements q_t an idea of total exchange area $\overline{G_i G_j}$, $\overline{G_i S_j}$, $\overline{S_i G_j}$, $\overline{S_i S_j}$ is used.

$$q_{t, g_i g_j} = \overline{G_i G_j} E_{g_i} = \overline{G_i G_j} \sigma T_{g_i}^4, \tag{2.34}$$

$$q_{t, g_i s_j} = \overline{G_i S_j} E_{g_i} = \overline{G_i S_j} \sigma T_{g_i}^4, \tag{2.35}$$

$$q_{t, s_i g_j} = \overline{S_i G_j} E_{s_i} = \overline{S_i G_j} \sigma T_{s_i}^4, \tag{2.36}$$

$$q_{t, s_i s_j} = \overline{S_i S_j} E_{s_i} = \overline{S_i S_j} \sigma T_{s_i}^4. \tag{2.37}$$

The subscript g and s represent the gas and wall elements, respectively, and i and j represent the emitting and absorbing elements, respectively; E denotes the blackbody emissive power of each element. For example, Eq. (2.34) calculates the total radiative energy emitted from gas element i and absorbed by gas element j. The total absorbed energy contains the energy emitted from the gas element i, attenuated while passing through the gas layer, and absorbed by the gas element j after reflected by surrounding walls once or several times. This occurs when the emissivity of the surrounding walls is below unity. The units of the variables q_t, σ, and T are watts, W/(m²K⁴), and Kelvin, respectively. The total exchange area $\overline{G_i G_j}$, $\overline{G_i S_j}$, $\overline{S_i G_j}$, and $\overline{S_i S_j}$ have a unit of area, square meters.

Once the values of the total exchange area are obtained, the temperature and the wall heat flux distributions in the system can be obtained by solving the following one set of energy equations:

Energy equation for gas element j—

$$\sum_i \overline{S_i G_j} E_{s_i} + \sum_i \overline{G_i G_j} E_{g_i} = 4K\Delta V E_{g_j} - q_{h, g_j} \tag{2.38}$$

or

Energy equation for wall element j—

$$\sum_i \overline{S_i S_j} E_{s_i} + \sum_i \overline{G_i S_j} = \varepsilon_j E_{s_j} - q_{a,s_j}. \qquad (2.39)$$

Left-hand terms represent the absorbed energy and right-hand terms represent the emitted or generated energy terms. The first terms on the LHS of the two equations denote the energy originally emitted from wall elements, and the second terms denote the energy from gas elements. The first terms on the RHS denote the self-emission of the element j. The variables q_h and q_a represent the heat generation by some chemical reactions in the gas element j and net heat load of the wall element j, respectively. The net heat load q_a means the removed heat from the wall element by some cooling devices to keep the temperature of the wall element constant. A part of the radiative energy emitted from the gas element j is absorbed by the element itself before the energy goes out of the element or once the energy is emitted out of the element and reflected back to the original element. That self-absorbed energy is represented by $\overline{G_j G_j} E_{g_j}$, which is included in the second term of the LHS of Eq. (2.38). The energy emitted from the wall element j and reflected back to and absorbed by the original element $\overline{S_j S_j} E_{s_j}$ is also included in the second term of the LHS of Eq. (2.39).

The procedure to solve radiative heat transfer by the zone method is as follows:

1. Obtain total exchange areas by the procedure mentioned later.
2. Give the values of heat generation, q_{h,g_j}, within each gas element.
3. Assign boundary conditions of wall elements. When temperature is given at each wall elements, the values of emissive power $E_{s_i} = (\sigma T_{s_i}^4)$ is given as the boundary condition. In this case, net heat load q_a is obtained for each wall element from the analysis. When net heat load is given as the boundary condition, the temperature is obtained for each element.
4. Solve a set of energy equations, (2.38) and (2.39).
5. Temperature of each gas element $T_{g_i}[= (E_{g_i}/\sigma)^{1/4}]$ and net wall heat load q_a or wall temperature $T_{s_i}[= (E_{s_i}/\sigma)^{1/4}]$ are obtained.

The total exchange areas are calculated from direct exchange areas $\overline{g_i g_j}$, $\overline{g_i s_j}$, $\overline{s_i g_j}$, $\overline{s_i s_j}$, which represent the radiative energy exchange between each element when all the wall elements in the system are assumed to be black (no reflection at wall). The direct exchange areas are defined by the

2.2. EXISTING METHODS OF SOLUTIONS

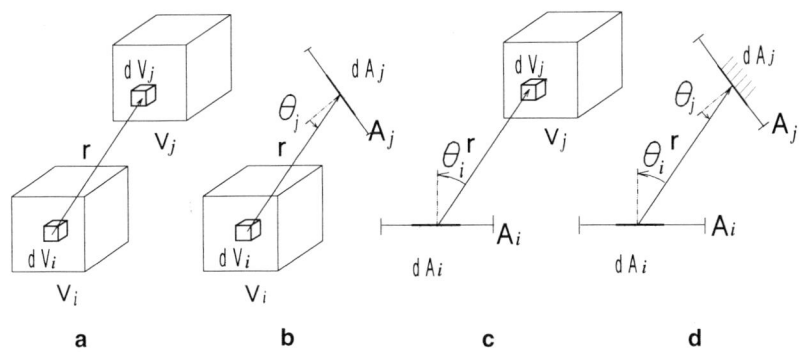

FIG. 2.8. Exchange between subdivided elements. Parts (a)–(d) values are obtained from Eqs. (2.44)–(2.47)

following equations:

$$q_{d, g_i g_j} = \overline{g_i g_j} E_{g_i} = \overline{g_i g_j} \sigma T_{g_i}^4, \tag{2.40}$$

$$q_{d, g_i s_j} = \overline{g_i s_j} E_{g_i} = \overline{g_i s_j} \sigma T_{g_i}^4, \tag{2.41}$$

$$q_{d, s_i g_j} = \overline{s_i g_j} E_{s_i} = \overline{s_i g_j} \sigma T_{s_i}^4, \tag{2.42}$$

$$q_{d, s_i s_j} = \overline{s_i s_j} E_{s_i} = \overline{s_i s_j} \sigma T_{s_i}^4. \tag{2.43}$$

The variable q_d represents the radiative energy, which is emitted from an element i, attenuated while passing through the gas layer, and absorbed by another element j. Though the direct exchange areas do not contain reflected energy, they have the same unit of square meters as the total exchange area. The values are obtained from the following equations for the system shown in Fig. 2.8:

$$\overline{g_i g_j} = \int_{V_i} \int_{V_j} \frac{K_i \, dV_i \, K_j \, dV_j \tau(r)}{\pi r^2}, \tag{2.44}$$

$$\overline{g_i s_j} = \int_{V_i} \int_{A_j} \frac{K_i \, dV_i \, dA_j \cos \theta_j \tau(r)}{\pi r^2}, \tag{2.45}$$

$$\overline{s_i g_j} = \int_{A_i} \int_{V_j} \frac{K_j \, dV_j \, dA_i \cos \theta_i \tau(r)}{\pi r^2}, \tag{2.46}$$

$$\overline{s_i s_j} = \int_{A_i} \int_{A_j} \frac{dA_i \cos \theta_i \, dA_j \cos \theta_j \tau(r)}{\pi r^2}. \tag{2.47}$$

Here, $\tau(r)$ is the transmittance of the gas layer with a thickness of r and is obtained from the following Beer's law:

$$\tau(r) = \exp\left[-\int_0^r K(r)\,dr\right]. \tag{2.48}$$

By comparing Eqs. (2.45) and (2.46), the following equation is obtained:

$$\overline{g_i s_j} = \overline{s_j g_i}. \tag{2.49}$$

As is shown in Eqs. (2.44)–(2.47), the direct exchange areas are the function of only the absorption coefficient of the gas elements K and the shape of the system. The method to obtain the values is shown later.

To derive the total exchange area from these direct exchange areas, radiosity W is incorporated. The radiosity denotes the total radiative energy flux emitted from each wall element. By using this radiosity, the total radiative energy emitted from a wall element i with an area A_i is expressed as

$$A_i W_i = A_i \varepsilon_i E_{s_i} + \rho_i \left(\sum_j \overline{s_j s_i} W_j + \sum_j \overline{g_j s_i} E_{g_j} \right). \tag{2.50}$$

The first term of the RHS represents the self-emission of the wall element, and the second term represents the reflected energy from this wall element. Here, ρ_i is the reflectivity of this wall element. The terms in parentheses are the incoming radiative energy onto the wall element. Equation (2.50) is made for each wall element. So, when direct exchange areas and the emissive power of each element E_{s_i} and E_{g_i} are given, a set of the Eq. (2.50) can be solved to obtain radiosities W_i of each wall element.

This radiosity is related to the total exchange area as follows. The total radiative energy emitted from a gas element i and absorbed by a wall element j is $\overline{G_i S_j} E_{g_i}$. So, the incident energy onto this wall element is $\overline{G_i S_j} E_{g_i}/\varepsilon_j$. Here, the radiosity of the wall element j is defined as $_{s_i} W_j$ when the emissive power of a gas element i, E_{g_i}, is assumed to be unity and the ones of all other gas and wall elements are set to be zero. Then, this radiosity is related to the total exchange area as follows:

$$A_j \cdot {}_{g_i}W_j = \frac{\overline{G_i S_j}}{\varepsilon_j} \rho_j. \tag{2.51}$$

By rearranging this equation, the total exchange area can be obtained

2.2. EXISTING METHODS OF SOLUTIONS

from the radiosity

$$\overline{G_i S_j} = \frac{A_j \varepsilon_j}{\rho_j} {}_{g_i}W_j. \tag{2.52}$$

For a uniform temperature field, the relation $E_{g_i} = E_{s_i} (= \sigma T^4)$ holds, and the net heat exchange between a gas element i and wall element j is zero:

$$\overline{G_i S_j} E_{g_i} = \overline{S_j G_i} E_{s_j} = 0. \tag{2.53}$$

Then, the following reciprocal relation between total exchange areas is derived:

$$\overline{G_i S_j} = \overline{S_j G_i}. \tag{2.54}$$

By setting the emissive power of a wall element i to be unity and the ones of all other elements to be zero, the total exchange area between wall elements is derived as follows:

$$\overline{S_i S_j} = \frac{A_j \varepsilon_j}{\rho_j} ({}_{g_i}W_j - \delta_{ij}\varepsilon_i). \tag{2.55}$$

Here, δ_{ij} is the delta function (and it equals 1 for $i = j$, and 0 for $i \neq j$). The total exchange area between gas elements is derived as follows by setting the emissive power of a gas element i to be unity and the ones of all other elements to be zero.

$$\overline{G_i G_j} = \overline{g_i g_j} + \sum_k \overline{s_k g_j} {}_{g_i}W_k. \tag{2.56}$$

The first term on the RHS represents the radiative energy emitted from the gas element i that reaches the gas element j directly. The second term represents the radiative energy originally emitted from the gas element i and reflected by wall element k that reaches the gas element j. Here, the direct exchange area $\overline{s_k g_j}$ is defined for the radiative energy emitted or reflected diffusely from the wall element k. So, a specularly reflecting wall cannot be treated by the zone method.

Then, the procedure to obtain the total exchange area is as follows:

1. Substitute $E_{g_i} = 1$, $E_{g_j} = 0$ (for $j \neq i$) and $E_{s_j} = 0$ to a set of Eq. (2.50), which is defined for each wall element, and obtain the values of W_j for each wall element. Then equate W_j to ${}_{g_i}W_j$.
2. Substitute $E_{s_i} = 1$, $E_{g_j} = 0$ and $E_{s_j} = 0$ (for $j \neq i$) to a set of Eq. (2.50), and obtain the values of W_j for each wall element. Then equate W_j and ${}_{s_i}W_j$.
3. Substitute ${}_{g_i}W_j$ or ${}_{s_i}W_j$ to Eqs. (2.52), (2.55) and (2.56) to obtain the total exchange area.

In conclusion, the radiative heat transfer can be analyzed by the zone method when a set of the direct exchange areas is obtained by carrying out the double integration in Eqs. (2.44) to (2.47). For elements with simple geometries, the direct exchange areas can be obtained from diagrams [23]. For elements with more complex shapes, the integration can be carried out by using the Monte Carlo method [24]. With this method, the direct exchange areas can be obtained numerically, even in three-dimensional systems. But, as is mentioned in the explanation of Eqs. (2.56), the zone method utilizes radiosity to derive total exchange area from direct exchange area, which requires the surface of the surrounding walls to be diffuse. So, specular reflecting walls cannot be treated. On the other hand, in the Monte Carlo analysis mentioned after Part II, the variables corresponding to the total exchange areas are directly obtained by the Monte Carlo technique. So, both diffuse and specular walls can be treated by the Monte Carlo method.

In the following, the same one-dimensional system treated in the previous section is solved by the zone method.

According to the textbook of Hottel and Sarofim [25], direct exchange areas $\overline{s_i s_j}$ is obtained as follows in the system shown in Fig. 2.9. From Eq. (2.47),

$$\overline{s_1 s_2} = \int_{A_1} dA_1 \int_{A_2} \frac{\cos^2 \theta \, dA_2 e^{-Kr}}{\pi r^2}, \tag{2.57}$$

where

$$r = L/\cos \theta,$$

$$dA_2 = 2\pi y \, dy = 2\pi (r \sin \theta) \frac{r d\theta}{\cos \theta} = 2\pi r^2 \tan \theta \, d\theta.$$

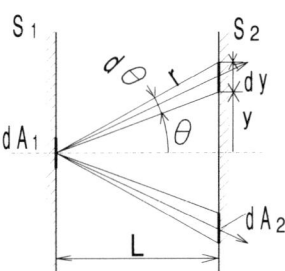

FIG. 2.9. Exchange between wall elements

2.2. EXISTING METHODS OF SOLUTIONS

Then,

$$\frac{\overline{s_1 s_2}}{A_1} = \int_0^{\pi/2} 2 \sin \theta \cos \theta e^{-kL/\cos \theta} \, d\theta. \tag{2.58}$$

Substituting $t = 1/\cos \theta$, and the direct exchange area $\overline{s_1 s_2}/A_1$, representing the energy transfer from the infinite flat plate 2 to a unit area on the flat plate 1, is newly defined as $\overline{s_1 s_2}$. Then, Eq. (2.58) can be rewritten as Eq. (2.59):

$$\overline{s_1 s_2} = \int_1^\infty \frac{2 e^{-KLt} \, dt}{t^3} = 2 E_3(KL). \tag{2.59}$$

Here, E_3 is the third exponential integral and the value is obtained from the summation of the following series:

$$E_3 \equiv \int_1^\infty \frac{e^{-rt}}{t^3} \, \partial t$$

$$= \frac{1}{2} - \tau + \frac{1}{2}\left(-0.577216 + \frac{3}{2} - \ln \tau\right)\tau^2$$

$$+ \sum_{m=3}^\infty \frac{(-\tau)^m}{(m-2)m!} \quad (\tau \neq 0).$$

$$= \frac{1}{2} \quad (\tau \neq 0) \tag{2.60}$$

Next, in the system shown in Fig. 2.10, the direct exchange area representing the energy transfer from an infinitely extending wall element to a gas element with a unit energy absorbing area can be obtained from the following equation:

$$\overline{s_1 g_j} = \overline{s_1 s_2} - \overline{s_1 s_3} = 2[E_3(Kx_{12}) - E_3(Kx_{13})]. \tag{2.61}$$

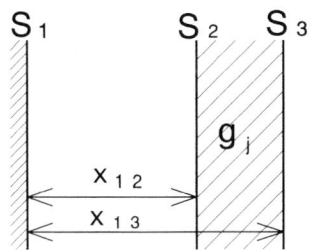

FIG. 2.10. Exchange between wall and gas elements

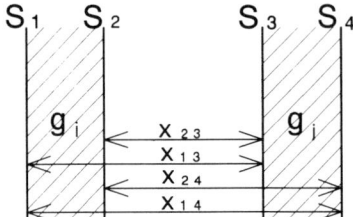

FIG. 2.11. Exchange between gas elements

This equation is derived from the fact that the absorbed energy in the gas element g_j is obtained from the difference between the energy that reaches the imaginary surfaces of s_2 and s_3. The direct exchange areas between gas elements are obtained using a similar idea, as shown in Fig. 2.11.

$$\overline{g_i g_j} = \overline{g_i s_3} - \overline{g_i s_4}$$
$$= \overline{s_3 g_i} - \overline{s_4 g_i}$$
$$= 2\{[E_3(Kx_{23}) - E_3(Kx_{13})] - [E_3(Kx_{24}) - E_3(Kx_{14})]\}. \quad (2.62)$$

The direct exchange area corresponding to self-absorption of gas element $\overline{g_i g_i}$ is obtained, in the system shown in Fig. 2.12, by subtracting the radiative energy escaping out of this element from the total self-radiation of this element:

$$\overline{g_i g_i} = 4Kx_{12} - \overline{g_i s_1} - \overline{g_i s_2}$$
$$= 4Kx_{12} - \overline{s_1 g_i} - \overline{s_2 g_i}$$
$$= 4Kx_{12} - 2[E_3(0) - E_3(Kx_{12})] - 2[E_3(0) - E_3(Kx_{12})]$$
$$= 4Kx_{12} - 2[1 - 2E_3(Kx_{12})]. \quad (2.63)$$

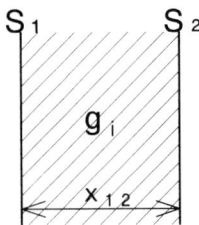

FIG. 2.12. Self-absorption in a gas element

2.2. EXISTING METHODS OF SOLUTIONS

```
1     ************************************************************
2     *                                                          *
3     *                          ZM                              *
4     *    1-D ANALYSIS ON RADIATION HEAT TRANSFER BY ZONING METHOD *
5     *                                                          *
6     ************************************************************
7           DIMENSION DSG(2,50),DGG(50,50),SG(2,50),GG(50,50),EG(50),TG(50)
8           open ( 6,file='PRN' )
9           write(*,100)
10    100   format(1h ,'input TW1 (K), TW2 (K), AKD, EM1, EM2, N'/)
11          READ(*,*) TW1,TW2,AKD,EM1,EM2,N
12          DAK=AKD/FLOAT(N)
13    *---------------------------------
14    * OBTAIN DIRECT INTERCHANGE AREAS
15    *---------------------------------
16          DSS=2.0*E3(AKD)
17          DSG(1,1)=1.0-2.0*E3(DAK)
18          DSG(2,N)=DSG(1,1)
19          DO 1000 I=2,N
20             DSG(1,I)=2.0*(E3(DAK*FLOAT(I-1))-E3(DAK*FLOAT(I)))
21    1000 CONTINUE
22          DO 1010 I=1,N-1
23             DSG(2,I)=2.0*(E3(DAK*FLOAT(N-I))-E3(DAK*FLOAT(N-I+1)))
24    1010 CONTINUE
25          DO 1020 I=1,N
26             DO 1030 J=1,N
27                IJD=ABS(I-J)
28                IF(IJD.EQ.0) THEN
29                   DGG(I,J)=4.0*(DAK-0.5+E3(DAK))
30                ELSE IF(IJD.EQ.1) THEN
31                   DGG(I,J)=1.0-4.0*E3(DAK)+2.0*E3(2.0*DAK)
32                ELSE
33                   DGG(I,J)=2.0*(E3(DAK*FLOAT(IJD-1))-2.0*E3(DAK*FLOAT(IJD))
34        1                      +E3(DAK*FLOAT(IJD+1)))
35                END IF
36    1030    CONTINUE
37    1020 CONTINUE
38    *---------------------------------
39    * OBTAIN TOTAL INTERCHANGE AREAS
40    *---------------------------------
41          R1=1.0-EM1
42          R2=1.0-EM2
43          D=1.0-R1*R2*DSS**2
44          S1S1=EM1**2*R2*DSS**2/D
45          S1S2=EM1*EM2*DSS/D
46          DO 1040 I=1,N
47             SG(1,I)=EM1*(DSG(1,I)+R2*DSS*DSG(2,I))/D
48    1040 CONTINUE
49          DO 1050 I=2,N
50             SG(2,I)=EM2*(DSG(2,I)+R1*DSS*DSG(1,I))/D
51    1050 CONTINUE
52          DO 1060 I=1,N
53             DO 1070 J=1,N
54                GG(I,J)=DGG(I,J)+(DSG(1,J)*R1*(DSG(1,I)+R2*DSS*DSG(2,I))
55        1                       +DSG(2,J)*R2*(DSG(2,I)+R1*DSS*DSG(1,I)))/D
56    1070    CONTINUE
57    1060 CONTINUE
58    *---------------------------------
59    * OBTAIN TEMPERATURE OF GAS ZONES
60    *---------------------------------
61          SBC=5.6687E-8
62          E1=SBC*TW1**4
63          E2=SBC*TW2**4
64          EG0=(E1+E2)*0.5
```

FIG. 2.13. Program of the zone method

```
 65         DO 1080 I=1,N
 66            EG(I)=EG0
 67   1080 CONTINUE
 68   2000 CONTINUE
 69         ERR=0.0
 70         DO 1090 I=1,N
 71            GGE=0.0
 72            DO 1100 J=1,N
 73               IF(I.NE.J) THEN
 74                  GGE=GGE+GG(J,I)*EG(J)
 75               END IF
 76   1100    CONTINUE
 77            EGI=(GGE+SG(1,I)*E1+SG(2,I)*E2)/(4.0*DAK-GG(I,I))
 78            ERRN=ABS(EGI-EG(I))
 79            IF(ERRN.GT.ERR) THEN
 80               ERR=ERRN
 81            END IF
 82            EG(I)=EGI
 83   1090 CONTINUE
 84         write(*,110) err
 85    110 format(1h ,'err=',e13.5)
 86         IF(ERR.GE.1.0E-5) GOTO 2000
 87         DO 1110 I=1,N
 88            TG(I)=(EG(I)/SBC)**0.25
 89   1110 CONTINUE
 90   *----------------------
 91   * OBTAIN WALL HEAT FLUX
 92   *----------------------
 93         GSE=0.0
 94         DO 1120 I=1,N
 95            GSE=GSE+SG(1,I)*EG(I)
 96   1120 CONTINUE
 97         QW=ABS((S1S1-EM1)*E1+S1S2*E2+GSE)
 98         QND=QW/(E1-E2)
 99   *----------------
100   * PRINT RESULTS
101   *----------------
102         WRITE(6,250)
103    250 FORMAT(1H ,'1-D ANALYSIS ON RADIATION HEAT TRANSFER BY ZONING METH
104        10D')
105         WRITE(6,300) TW1,TW2,AKD,EM1,EM2,N
106    300 FORMAT(1H ,'TW1=',E12.5,'(K)',2X,'TW2=',E12.5,'(K)'/
107        1 ' OPTICAL THICKNESS=',E12.5,2X,'EM1=',E12.5,2X,'EM2=',
108        2 E12.5/' NUMBER OF GAS ELEMENTS=',I3/)
109         DO 1130 I=1,N
110            WRITE(6,400) I,TG(I)
111    400    FORMAT(1H ,3X,'I=',I3,2X,'T=',E12.5,'(K)')
112   1130 CONTINUE
113         WRITE(6,500) QW,QND
114    500 FORMAT(1H ,'QW=',E12.5,'(W/m2)',4X,'QND=',E12.5)
115         STOP
116         END
117   *
118         FUNCTION E3(X)
119   *-----------------------------------------------
120   * OBTAIN THE VALUE OF 3RD EXPONENTIAL INTEGRAL
121   *-----------------------------------------------
122         XM=X**2
123         E3=0.5-X+0.5*(-0.577216+1.5-ALOG(X))*XM
124         IFAC=2
125         DO 10 M=3,20
126            IFAC=IFAC*M
127            XM=XM*(-X)
128            E3=E3-XM/FLOAT((M-2)*IFAC)
129     10 CONTINUE
130         RETURN
131         END
```

FIG. 2.13. (*Continued*)

2.2. EXISTING METHODS OF SOLUTIONS

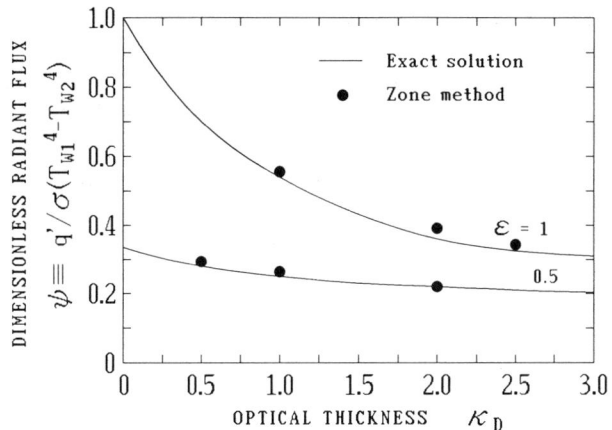

FIG. 2.14. A comparison of the results obtained by the zone method and the exact solution

By substituting these direct exchange areas into Eq. (2.50), $_{g_i}W_j$ and $_{s_i}W_j$ are obtained by the procedure mentioned before. The total exchange areas for a unit energy receiving area in a one-dimensional system are obtained by substituting these variables into Eqs. (2.52) to (2.56):

$$\overline{S_1S_1} = \frac{\varepsilon_1^2 \rho_2 \overline{s_1s_2}^2}{D}, \qquad (2.64)$$

$$\overline{S_1S_2} = \frac{\varepsilon_1 \varepsilon_2 \overline{s_1s_2}}{D}, \qquad (2.65)$$

$$\overline{S_1G_i} = \frac{\varepsilon_1\left(\overline{s_1g_i} + \rho_2\overline{s_1s_2}\cdot\overline{s_2g_i}\right)}{D}, \qquad (2.66)$$

$$\overline{G_iG_j} = \overline{g_ig_j} + \frac{\overline{s_1g_j}\rho_1\left(\overline{s_1g_i} + \rho_2\overline{s_1s_2}\cdot\overline{s_2g_i}\right) + \overline{s_2g_j}\rho_2\left(\overline{s_2g_i} + \rho_1\overline{s_1s_2}\cdot\overline{s_1g_i}\right)}{D}. \qquad (2.67)$$

Here,

$$D = 1 - \rho_1\rho_2\overline{s_1s_2}^2. \qquad (2.68)$$

By using these total exchange areas and giving the wall temperatures of the bounding two parallel walls, the temperature of each gas element between the walls can be calculated from the set of energy equations for the gas elements, Eq. (2.38). In the present analysis, the internal heat generation in the gas element is set to be zero ($q_{h,g_i} = 0$). Wall heat flux, q_{a,s_j} can be obtained by substituting these results into Eq. (2.39).

The FORTRAN program in Fig. 2.13 shows these procedures. The input variables are wall temperatures of the bounding walls TW1 and TW2, optical distance between the walls AKD, emissivities of the walls EM1 and EM2, and the number of gas elements N. The format of the output is the same as the program for two-flux model shown in Figs. 2.4 and 2.5. Figure 2.14 shows the results of the program, which fits well with the exact solution.

Part II
PRINCIPLES OF MONTE CARLO METHODS

Chapter 3
Formulation

3.1. Introduction

In the study of a combined radiative–convective–conductive heat transfer process, the divergent of the radiative heat flux $\mathbf{q}_r$ in the energy balance equation, Eq. (2.1), must be known at each location in the system. Its magnitude can be determined by Eq. (2.4). The scattering decay, $\sigma_{s\lambda}$ in the second term on the RHS of Eq. (2.4), is equal to and cancels with the emission due to scattering, the third term. Hence, the radiative contribution is determined by evaluating the heat loss through self-emission from the differential volume

$$4\pi \int_0^\infty K_\lambda(\lambda,T) I'_{\lambda b}(\lambda)\, d\lambda$$

and the heat gain of the differential volume by absorbing radiant energy

$$4\pi \int_0^\infty [K_\lambda(\lambda, T) + \sigma a(\lambda)] \bar{I}_\lambda(\lambda)\, d\lambda.$$

Both the heat loss by emission and the heat gain by absorption are spatially dependent, but directionally independent, scalar quantities. The former, which is proportional to the fourth power of the local temperature, can be evaluated rather easily. Therefore, radiative heat transfer analysis would be simplified if the latter could be easily determined. The conventional method is to determine simultaneously all radiative heat gains in the entire system and thus define the directionally dependent source function and radiation intensity at all locations in order to solve Eqs. (2.6) and (2.9).

Physically, energy transport by radiation is a combination of scattering and absorption of a large number of independent photons, which are issued from either gas molecules or solid wall molecules. The source function and radiation intensity are used only as a means of expressing the macroscopic behavior of a radiative process. In contrast, the Monte Carlo method directly simulates the physical process of radiative heat transport. It traces and collects the scattering and absorption behavior of a large number of independent radiative energy particles, namely, photons, which are emitted from each point in the system. Numerical computations are then performed to determine the incident radiant heat from the surroundings to the system and the absorption distribution of self-radiant heat emitted from each position within the system. We can evaluate the distribution of incident radiation that is absorbed by each differential volume, within the second term on the RHS of Eq. (2.4). Now, various postulations and restrictions are eliminated because it becomes unnecessary to determine the direction-dependent source function and the radiation intensity. The radiative heat transfer analysis is greatly simplified through the use of the Monte Carlo method.

It has been stated that the Monte Carlo method traces the behavior of each photon. However, the tracing of all photons requires an enormous amount of computational time and computer memory. Instead of a large number of photons, the method traces the behavior of a randomly selected, finite number of energy particles. The defects of the conventional Monte Carlo method include data scattering, resulting from a reduction in the number of selected samples, and an increase in the computational time required for tracing a large number of energy particles. This problem has been solved with the advent of high-speed computers and the introduction of the READ (radiant energy absorption distribution) technique [6, 7] to be presented in Chapter 4. The successive relaxation method utilized the energy exchange between adjacent elements in the numerical integration of combined conductive–convective heat transfer equations for temperature distribution. The modified Monte Carlo method using the READ technique can determine the temperature distribution by treating the energy exchange between the elements that are adjacent or remotely separated.

This chapter presents the approach, fundamental equations, and examples of the modified Monte Carlo method being employed to analyze combined radiative–convective heat transfer problems in the systems that consist of radiative–absorptive gray gases and gray walls. Such problems have broad practical applications. By using the modified Monte Carlo method, the temperature distribution in the gas phase as well as the temperature and heat flux distributions of the wall can be determined by specifying the geometry of the system and operating conditions in the program RADIAN, which is discussed in Chapter 6.

3.2. Heat Balance Equations for Gas Volume and Solid Walls

Consider a combined heat transfer process in a gas–wall system (with the gas including flames and combustion gases). The gas volume and solid

3.2. HEAT BALANCE EQUATIONS FOR GAS VOLUME AND SOLID WALLS

wall are subdivided into an appropriate number of elements, as depicted in Fig. 3.1. The heat balance equation for an element is equivalent to that resulting from an integration of the heat equation, Eq. (2.1).

The heat balance equations for the gas and wall elements can be expressed as follows:

As a gas element:

$$q_{r,\text{out},g} + q_{c,gw} + q_{f,\text{out}} = q_{r,\text{in},g} + q_{h,g} + q_{f,\text{in}} \tag{3.1}$$

or wall element:

$$q_{r,\text{out},w} + q_a = q_{r,\text{in},w} + q_{c,gw}. \tag{3.2}$$

In both equations, the heat flow components for flowing out of the element are placed on the LHS, with those components flowing into the element on the RHS. The q_a term in Eq. (3.2) denotes the net heat flow rate received by the wall element. It is placed on the LHS being equal to heat removal from the wall element by cooling under an equilibrium condition. The expressions for obtaining each term in Eqs. (3.1) and (3.2) are described next.

The radiative energy emitted from a gas element of volume ΔV is derived from Eq. (1.32) as

$$q_{r,\text{out},g} = 4(1 - \alpha_{s,g})\sigma K T_g^4 / \Delta V, \tag{3.3}$$

in which $\alpha_{s,g}$ represents the self-absorption ratio of the radiative energy being emitted from the element. In the derivation of Eq. (1.32), consideration is given to a gas element of small volume from which the entire emissive heat goes into the surroundings. However, for a gas element of

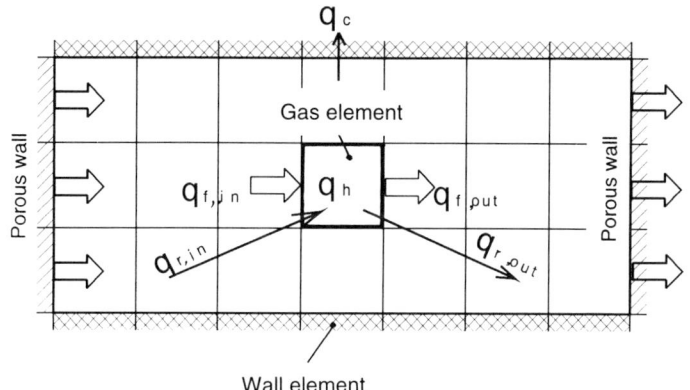

FIG. 3.1. Analytical model

finite volume, it is necessary to take into account the fraction of emissive heat that is absorbed internally by the element itself (self-absorption). For convenience in analysis, self-absorption includes the fractions being reflected by the wall and scattered by the gas medium that have returned to, and been absorbed by, the originating element. The magnitude of $\alpha_{s,g}$ is obtained in the course of calculating the READ value by means of the Monte Carlo method, as mentioned later in Section 4.2.

The radiative energy emitted from a wall element of a wetted area ΔS is derived from Eq. (1.6) as

$$q_{r,\text{out},w} = (1 - \alpha_{s,w}) \varepsilon \sigma T_w^4 \Delta S. \tag{3.4}$$

Here, $\alpha_{s,w}$ signifies the self-absorption ratio by the wall element. Analogous to $\alpha_{s,g}$ in Eq. (3.3), it represents the fraction of the radiative heat emitted from the wall element that returns to, and is absorbed by, the original element through reflection and scattering.

The incident radiation on each element is a summation of the radiative heat components from all elements:

$$q_{r,\text{in},i} = \underset{\text{gas}}{\sum R_d \cdot q_{r,\text{out},g}} + \underset{\text{wall}}{\sum R_d \cdot q_{r,\text{out},w}} \quad \text{for } (i = g, w). \tag{3.5}$$

Here, R_d is the fraction of radiative energy emitted from all gas and wall elements excluding the element under consideration, and is absorbed by the element. The R_d's are a set of constants in the READ technique and their magnitude is determined by means of the Monte Carlo method.

The convective heat transfer from gas elements to wall elements is determined by

$$q_{c,gw} = h_{gw}(T_g - T_w) \Delta S. \tag{3.6}$$

Here, h_{gw} is the convective heat transfer coefficient, which can be obtained by theoretical analysis, experiment, or from handbooks, etc.

The fluid entering a gas element carried the enthalpy $q_{f,\text{in}}$ and is evaluated by the expression

$$q_{f,\text{in}} = W_{g,\text{up}} C_p T_{g,\text{up}} \Delta S_g, \tag{3.7}$$

where W_g represents the rate of mass flowing through each surface of the gas element, ΔS_g. Its distribution is determined separately by a finite difference method, finite element method, experiment, etc.

Similarly, the enthalpy leaving a gas element is

$$q_{f,\text{out}} = W_g C_p T_g \Delta S_g. \tag{3.8}$$

The preceding expressions, Eqs. (3.3)–(3.8), form the basic equations for solving the combined radiation–convection heat transfer problems. When all physical properties, elements, and system geometry, $\alpha_{s,g}$, $\alpha_{s,w}$, R_d, h_{gw}, W_g, and T_w as a boundary condition or q_a are specified, Eqs. (3.1) and (3.2), upon the substitution of Eqs. (3.3)–(3.8), become a function of only the temperature of each element. Therefore, if Eqs. (3.1) and (3.2) for each element are solved by the successive relaxation method, one can obtain the distributions of the gas temperature T_g and the wall heat flux q_a or the wall temperature T_w. Among the parameters mentioned earlier that are needed for analysis, physical properties, geometry of the element and system, h_{g-w}, W_g, and boundary conditions can be specified as the input data, whereas the remaining parameters $\alpha_{s,g}$, $\alpha_{s,w}$, and R_d must be evaluated using the Monte Carlo method. These three variables, determined by the Monte Carlo method, are expressed in the normalized form, by dividing the absorption distribution of radiative energy emitted from each element in the system by the total emitted energy from the element. Hence, their magnitude is not affected by the absolute value of radiation emission from each element. In other words, as long as the geometry of the elements and system is specified and the radiative properties are independent of temperature, these variables may be treated as constants during the computational process of temperature convergence. They need to be calculated only once prior to the computation of temperature convergence, using the Monte Carlo method, which is presented in the next chapter. In the presence of heat conduction due to molecular diffusion or turbulent eddy diffusion on a jet surface with shear forces, heat transfer becomes appreciable in the direction perpendicular to the fluid flow. Under such a circumstance, the heat transfer between the elements induced by molecular diffusion or turbulent eddy diffusion should be incorporated into Eq. (3.1) in a form similar to $q_{f,\text{in}}$ and $q_{f,\text{out}}$.

In the case where physical properties depend strongly on temperature, the radiative heat transfer computation for $\alpha_{s,g}$, $\alpha_{s,w}$, and R_d by means of the Monte Carlo method together with the energy equations (3.1) and (3.2) must be solved simultaneously for T_g, q_a, and T_w by an iterative procedure.

3.3. Simulation of Radiative Heat Transfer

The task of simulating radiative heat transfer consists of four parts: simulations of gas absorption, emission from gas volumes, emission from solid walls, and reflection and absorption by solid walls.

3.3.1. SIMULATION OF GAS ABSORPTION

In solving radiative heat transfer problems by means of the Monte Carlo method, radiative energy is not treated as a continuous, variable entity, but is considered a collection of photons, each with a fixed amount of energy ($h\nu$). Radiative energy which is considered a lumped quantity, that is, continuum, in the conventional approach, is treated as a distributed quantity (i.e., the number of photons) in the Monte Carlo approach. Through this distributed approach, the computation of radiative heat exchanges among each element in the system is not done by integrating continuums, that is, radiative energy intensities, but by means of summing the behavior of each energy particle.

To demonstrate an actual example of how to treat radiative heat transfer using the Monte Carlo method, an attenuation of radiative energy in the gas is described as follows.

Radiative energy of intensity I_0, which is emitted from point P in the x direction, attenuates continuously, as shown in Fig. 3.2, in the gas through absorption according to Beer's law, Eq. (1.25). The energy being absorbed in the band of dx at a distance x from point P is $I_0 K e^{-Kx} dx$ or, equivalently, $I_0 K e^{-Kx}$ per unit distance in the x direction. Let us treat this problem by means of the Monte Carlo method. Consider an ejection of N particles, each having the energy of I_0/N, in the x direction. The number N is quite arbitrary, but a large N would result in better accuracy with the penalty of a longer computational time. It is postulated that (a) all radiative energy of each particle will eventually be absorbed by gas molecules at a certain location x; and (b) the energy possessed by each particle remains unchanged during its flight. This is precisely the same concept as the transfer of radiative energy by photons. It is important, therefore, to determine the flight distance of each of the N particles so

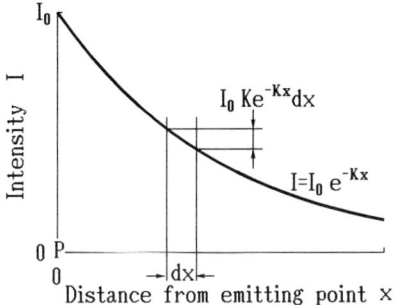

FIG. 3.2. Attenuation of radiative energy in a gas layer

3.3. SIMULATION OF RADIATIVE HEAT TRANSFER

that the absorption distribution of radiative energy being carried by the N particles on the x axis is equal to that being evaluated using Eq. (1.25). Thus, one can simulate the attenuation of radiative energy in compliance with Beer's law.

The inverse transformation method [26] is one technique to determine the number array η (flight distance of a particle in the present case) that fits a certain probable density distribution, $f(\eta)$. The transformation equation is

$$\xi = \int_{-\infty}^{\eta} f(y) \, dy. \tag{3.9}$$

Here, ξ is the variable that makes the value of η in the range of 0 to 1 to be equal to 1. When the variable ξ is replaced by the uniform random variable R, Eq. (3.9) reads

$$R = \int_{-\infty}^{\eta} f(y) \, dy. \tag{3.10}$$

The number array $\{\eta_i\}$, which is obtained by substituting the uniform random numbers [0, 1] for R in this equation, would become a random number array possessing the probability density distribution $f(\eta)$.

From the N energy particles ejected at $x = 0$, the radiative energy being absorbed over the flight distance between S and $S + dS$ is $I_0 K e^{-KS} dS$, as mentioned earlier. Because each particle has energy I_0/N, the number of particles being absorbed in the band of dS is

$$\frac{I_0 K e^{-KS} dS}{I_0/N} = NKe^{-KS} dS. \tag{3.11}$$

The probability density distribution of those energy particles having a flight distance of S is

$$f(S) \, dS = NKe^{-KS} \, dS/N = Ke^{-KS} \, dS. \tag{3.12}$$

Because $S \geq 0$, a combination of Eqs. (3.9) and (3.12) yields

$$\xi = \int_{-\infty}^{S} f(y) \, dy = \int_{0}^{\infty} f(y) \, dy = \int_{0}^{S} Ke^{-KY} \, dy = 1 - e^{-KS}. \tag{3.13}$$

An application of Beer's law on the RHS of Eq. (3.13) results in

$$1 - e^{-KS} = \frac{I_0 - I}{I_0} = 1 - \frac{I}{I_0}. \tag{3.14}$$

This expresses the probability (absorption probability) that radiative energy emitted from point P (at $x = 0$) will be absorbed over the distance from the origin $x = 0$ to $x = S$. The RHS of this probability equation, Eq. (3.13), is equated to the uniform random number between 0 and 1:

$$R_s = 1 - e^{-KS}. \tag{3.15}$$

The equation is rewritten as

$$KS = -\ln(1 - R_s), \tag{3.16}$$

where R_s is the uniform random number to be employed for the determination of S. If the value of KS obtained by the substitution of the [0, 1] uniform random number into Eq. (3.16) is taken as the flight distance (namely, absorption distance) of the N energy particles, then the absorption distribution of these particles would satisfy Beer's law. The flight distance KS of each particle can be determined from Fig. 3.3 by reading the value of KS on the abscissa (for example [0, b]), which corresponds to a specified [0, 1] uniform random number on the ordinate (i.e., [0, a]).

If the absorption coefficient of gas in system K is uniform, the flight distance of each particle would be $KS / K = S$. In the case of a system with K varying with each gas element, the gas elements are numbered 1, 2, ..., along the flight passage of energy particles; their absorption coefficients, K_i; and the flight distances of energy particles through the elements, S_i. For the particle being absorbed by the n'th gas element, n must satisfy the relation

$$\sum_{i=1}^{n-1} K_i S_i \leq KS \leq \sum_{i=1}^{n} K_i S_i, \tag{3.17}$$

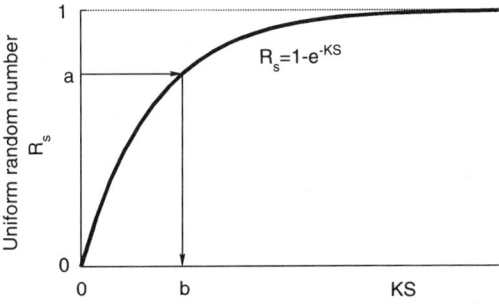

FIG. 3.3. Principle to determine flight distance of each energy particle

3.3. SIMULATION OF RADIATIVE HEAT TRANSFER

where KS is from Eq. (3.16). The value of n is then employed to determine the flight distance as

$$S = \sum_{i=1}^{n-1} S_i + \frac{\left(KS - \sum_{i=1}^{n-1} K_i S_i\right)}{K_n}. \tag{3.18}$$

The first term on the RHS of the equation indicates the total passage length of a particle through the gas elements from $i = 1$ to $(n - 1)$; the second term is the flight distance through the n'th element before absorption. In the radiative heat transfer analysis using the Monte Carlo method, one has only to identify the gas element that absorbs the particle. It is unnecessary to know the exact location where the particle is absorbed by the element. Therefore, Eq. (3.18) is not employed in the actual analysis; instead, Eq. (3.17) determines the absorbing element of each energy particle.

Figure 3.4 presents a list of the program BEER, which uses Eq. (3.16) to simulate Beer's law on the attenuation of radiative energy in an absorbing gas, by means of the Monte Carlo method. Line 15 in the program reads in the number of energy particles, NRAY, and the gas absorption coefficient, AK (unit of inverse meters). The flight distance S of each particle is determined in the statement between lines 16 through 40, with its histogram constructed. Line 18 is Eq. (3.16). To construct the histogram of the flight distance of each particle, a length of 10 m from the origin (i.e., $x = 0$ to 10 m) is subdivided into 1-m segments, and the number of particles that enters each flight segment is calculated in the statement between lines 19 through 39. The uniform random number RAN to be used in Eq. (3.16) is determined by the subroutine RANDOM, lines 63 through 68. This subroutine, utilizing the congruential method, is a program that begins with the variable RAND in lines 9 and 11 and eventually generates an array of uniform random numbers, RAN. It is similar to other routines for generating random numbers in other programs given in the text. At every CALL, a new value of random numbers enters into the variable RAN and the subroutine repeats its function. The statement following line 41 concerns the summary and printout of the calculated results. Line 48 determines ABSORP, the fraction of radiative heat being adsorbed in each of the 1-m-wide segments (out of the total radiative heat entering the gas layer). It is equal to the number of energy particles $N(J)$ with their flight distance S reaching a segment J (equal to the number of particles being absorbed in the segment), divided by the total number of energy particles, NRAY. In line 56, Beer's law [Eq. (1.25)] is applied to evaluate ABSPR, the amount of radiative energy of unit strength being

```
 1    ************************************************************************
 2    *                                                                      *
 3    *                              BEER                                    *
 4    *               MONTE CARLO SIMULATION OF BEER'S LAW                   *
 5    *                                                                      *
 6    ************************************************************************
 7          DIMENSION N(10)
 8          CHARACTER*1 H(80),STAR,BLANK
 9          REAL*8 RAND
10          DATA STAR/'*'/,BLANK/' '/,N/10*0/
11          RAND=5249347.0D0
12          open ( 6,file='PRN' )
13          write(*,100)
14    100 format(1h ,'input (nmax) and (absorption coeff.)')
15          READ(*,*) NRAY,AK
16          DO 1000 I=1,NRAY
17             CALL RANDOM(RAN,RAND)
18             S=-ALOG(1.0-RAN)/AK
19             IF (S.LT.1.) THEN
20                N(1)=N(1)+1
21             ELSE IF (S.LT.2.) THEN
22                N(2)=N(2)+1
23             ELSE IF (S.LT.3.) THEN
24                N(3)=N(3)+1
25             ELSE IF (S.LT.4.) THEN
26                N(4)=N(4)+1
27             ELSE IF (S.LT.5.) THEN
28                N(5)=N(5)+1
29             ELSE IF (S.LT.6.) THEN
30                N(6)=N(6)+1
31             ELSE IF (S.LT.7.) THEN
32                N(7)=N(7)+1
33             ELSE IF (S.LT.8.) THEN
34                N(8)=N(8)+1
35             ELSE IF (S.LT.9.) THEN
36                N(9)=N(9)+1
37             ELSE IF (S.LT.10.) THEN
38                N(10)=N(10)+1
39             END IF
40    1000 CONTINUE
41          WRITE(6,200) NRAY,AK
42    200 FORMAT(1H ,'MONTE CARLO SIMULATION OF BEER''S LAW'/8X,'NMAX=',I6,4
43         *X,'ABSORPTION COEFF.=',F5.2,'(1/M)'/)
44          WRITE(6,300)
45    300 FORMAT(1H ,'DISTANCE',2X,'EXACT ABSORPTION',5X,'CALCULATED ABSORPT
46         *ION')
47          DO 2000 J=1,10
48             ABSORP=FLOAT(N(J))/FLOAT(NRAY)
49             NN=IFIX(80.0*ABSORP)
50             DO 3000 I=1,NN
51                H(I)=STAR
52    3000    CONTINUE
53             DO 4000 I=NN+1,80
54                H(I)=BLANK
55    4000    CONTINUE
56             ABSPR=EXP(-AK*FLOAT(J-1))-EXP(-AK*FLOAT(J))
57             WRITE(6,400) J,ABSPR,ABSORP,(H(I),I=1,80)
58    400     FORMAT(1H ,3X,I2,7X,E12.5,3X,E12.5,4X,80A1)
59    2000 CONTINUE
60          STOP
61          END
62    *
63          SUBROUTINE RANDOM(RAN,RAND)
64          REAL*8 RAND
65          RAND=DMOD(RAND*131075.0D0,2147483649.0D0)
66          RAN=SNGL(RAND/2147483649.0D0)
67          RETURN
68          END
```

FIG. 3.4. Program to simulate Beer's law

3.3. SIMULATION OF RADIATIVE HEAT TRANSFER

a MONTE CARLO SIMULATION OF BEER'S LAW
         NMAX= 50000   ABSORPTION COEFF.= .10(1/M)

DISTANCE	EXACT ABSORPTION	CALCULATED ABSORPTION	
1	.95163E-01	.93280E-01	*******
2	.86107E-01	.85520E-01	******
3	.77913E-01	.78440E-01	******
4	.70498E-01	.71900E-01	*****
5	.63789E-01	.64900E-01	*****
6	.57719E-01	.57620E-01	****
7	.52226E-01	.51860E-01	****
8	.47256E-01	.48160E-01	***
9	.42759E-01	.43240E-01	***
10	.38690E-01	.38140E-01	***

b MONTE CARLO SIMULATION OF BEER'S LAW
         NMAX= 50000   ABSORPTION COEFF.= .50(1/M)

DISTANCE	EXACT ABSORPTION	CALCULATED ABSORPTION	
1	.39347E+00	.39404E+00	*******************************
2	.23865E+00	.23902E+00	*******************
3	.14475E+00	.14366E+00	************
4	.87795E-01	.88660E-01	*******
5	.53250E-01	.52940E-01	****
6	.32298E-01	.33240E-01	**
7	.19590E-01	.18600E-01	*
8	.11882E-01	.11820E-01	
9	.72066E-02	.69000E-02	
10	.43710E-02	.43400E-02	

FIG. 3.5. (a) Output of BEER program, example 1; (b) Output of BEER program, example 2

absorbed in the interval between $(J - 1)$ m and J m during its transmission through the gas layer. The statement from lines 49 through 55 calculates the amount of radiative heat being absorbed in every 1-m segment for a presentation in bar-graph form. The result shows the absorption distribution based on 80 stars to represent the strength of radiative energy incident on the gas layer.

Figures 3.5(a) and (b) depict application examples of the computer program. Figure 3.5 corresponds to the absorption coefficient of 0.1 m^{-1} with 50,000 energy particles. DISTANCE signifies distance in a gas layer (in meters); and EXACT ABSORPTION is the exact solution for the absorption in each of the 1-m intervals obtained directly from Beer's law. They are expressed in bar-graph form. It is disclosed from the program, through the use of 50,000 particles, that the errors in the solution obtained by the Monte Carlo method are within 1 to 2% of the exact solution. Figure 3.6 shows the relationship between the particle number used in the analysis by the Monte Carlo method and the convergence of its solution to the exact one. This figure plots the number of energy particles converted from the fraction of absorbed radiative energy in the 1-m segment next to

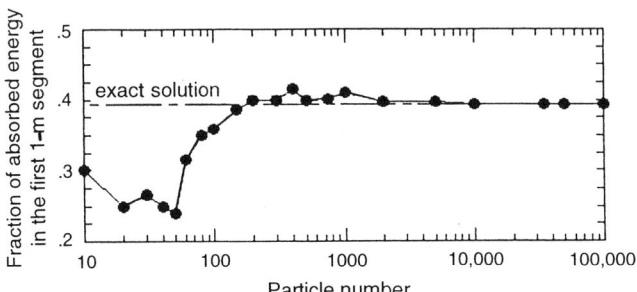

FIG. 3.6. Convergence of Monte Carlo calculations

the entrance of the gas layer obtained by the Monte Carlo method. The figure shows that the errors are reduced to about 4% in the case of 1000 particles and converge to within 1 to 2% for particle numbers exceeding 10,000. Figure 3.5(b) presents the analytical results of the same problem as Figure 3.5(a), but the gas absorption coefficient has been increased by 0.5 m^{-1}. Note that the radiative energy absorption in the entrance region of the gas layer is enhanced with an increase in the gas absorptive coefficient.

3.3.2. Simulation of Radiation Emission from Gas Volume

3.3.2.1. *Direction of Ejection of Energy Particles from Gas Elements*

A simulation of the behavior of radiative energy emitted from a gas volume, by means of the Monte Carlo method, is achieved by tracing the paths of a certain number (a very large number) of energy particles that are ejected from the gas volume into its surroundings. When this radiative energy is divided into a large number of energy particles, the flight distance of each particle can be determined by Eq. (3.16). As long as the propagation of radiative energy is not a one-dimensional beam, it is necessary to determine the direction of ejection of each particle. Even if the system is one- or two-dimensional, the radiative energy emitted from a gas volume within the system will propagate three dimensionally in the surroundings. Therefore, it is imperative to trace the radiative energy particles three dimensionally. Because gas radiation is diffuse, a uniform random number is employed to adjust the ejection direction of energy particles uniformly in the surroundings of the ejection source, that is, the gas volume. In the spherical coordinates shown in Fig. 3.7, since the particles are ejected uniformly within the $0-2\pi$ space in the θ direction,

3.3. SIMULATION OF RADIATIVE HEAT TRANSFER

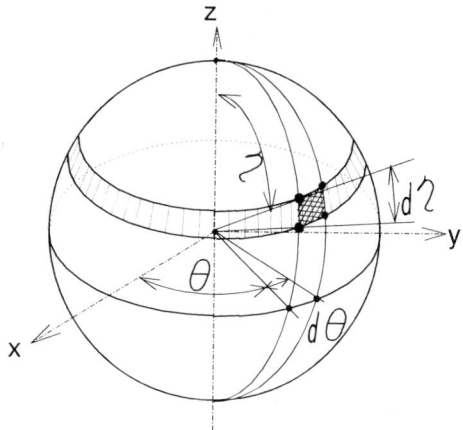

FIG. 3.7. Angles of radiative energy emission from a gas element

the ejection angle of each particle in the θ direction can be determined using the uniform random number

$$R_\theta = \theta/2\pi. \tag{3.19}$$

Next we consider the η direction. Let $f(\eta)$ be the probability density distribution for the ejection direction of each particle and $f(\eta)\,d\eta$ be the probability for an ejection in the angle between η and $\eta + d\eta$. Diffuse radiation implies that the radiative energy passing through a unit surface area of a sphere centered at the ejection source, that is, the gas volume, is uniform everywhere on the spherical surface. Hence, the probability is equal to the cross-hatched surface area in Fig. 3.7 divided by the total surface area:

$$f(\eta)\,d\eta = (2\pi r \sin \eta) r \, d\eta/(4\pi r^2) = (1/2)\sin \eta\,d\eta. \tag{3.20}$$

Taking into account that $\eta \geq 0$, an application of the inverse transformation method yields

$$\xi = \int_0^\eta (1/2)\sin \eta\, d\eta = (1 - \cos \eta)/2. \tag{3.21}$$

This equation expresses the probability that energy will be emitted in the angle between 0 and η, out of the entire emitted energy. Like the other accumulative absorption probability, the present one takes a value between 0 and 1 as a uniform probability in accordance with the inverse transformation method. Hence, it can be replaced by the uniform random number R_η to determine the ejection angle of each energy particle.

$$R_\eta = (1 - \cos \eta)/2. \tag{3.22}$$

We can now utilize the two uniform random numbers R_θ and R_η to determine the ejection directions (θ, η) of each energy particle as

$$\theta = 2\pi R_\theta, \tag{3.23}$$

$$\eta = \cos^{-1}(1 - 2R_\eta). \tag{3.24}$$

3.3.2.2. Location of Ejection of Energy Particles inside Gas Elements

As mentioned in Section 3.2, the Monte Carlo method divides the system into a finite number of gas elements and their surrounding wall elements, and then analyzes radiative energy exchanges among them. It is allowed to consider all radiative energy emitted from the central point on the surface of a wall element, just like other numerical methods using various mesh divisions. As for the gas element, if the radiative energy is emitted only from the element center, the average value of the absorbed energy by the element itself, which is calculated along the passage from the interior to the exterior of the element, would be higher than that in the actual case (uniformly emitted from all locations in an element having uniform interior temperature). Consequently, self-absorption—the fraction of absorption within its own element—increases, while the radiative energy exchange among different elements diminishes. This effect becomes significant with an increase in element size. To prevent this in the use of the Monte Carlo method, it is necessary to consider a uniform energy emission from all locations within the gas element, as is the situation in the actual phenomenon. If the gas elements are cubic and brick-shaped, the ejecting position of each energy particle (x_0, y_0, z_0) can be described using three uniform random numbers, R_1, R_2, and R_3, as

$$x_0 = (R_1 - 0.5)\Delta x + x_c, \tag{3.25}$$

$$y_0 = (R_2 - 0.5)\Delta y + y_c, \tag{3.26}$$

$$z_0 = (R_3 - 0.5)\Delta z + z_c. \tag{3.27}$$

Here, (x_c, y_c, z_c) denotes the coordinates of the element center and Δx, Δy, and Δz represent the lengths of a brick-shaped body in the x, y, and z directions, respectively. If the gas elements are not a brick-shaped body, the ejecting points of energy particles can be uniformly distributed within the elements using a number of uniform random numbers.

3.3. SIMULATION OF RADIATIVE HEAT TRANSFER

In summary, if the Monte Carlo method is to be used to simulate the emission of radiative energy from the gas elements, Eqs. (3.25)–(3.27) must be utilized to describe the starting point of multiple energy particles within the element, Eqs. (3.23) and (3.24) for the ejecting direction, and Eq. (3.16) for the flight distance. As a result, the analysis determines the locus of each energy particle, described by six uniform random numbers. The terminal point of the locus of each particle serves as the absorption point for radiative energy, which is stored in the computer. The absorption distribution of all radiative energy emitted from the gas elements can be obtained from the distribution of terminal points of these loci in the system.

3.3.3. SIMULATION OF RADIATION EMISSION FROM SOLID WALLS

To simulate the emission of radiative energy from the wall element by the Monte Carlo method, a fixed number of energy particles are uniformly ejected from the entire surface of the element; the distribution of the ejecting angles of all emitted energy particles must obey Lambert's cosine law, Eq. (1.7); and the flight distance of each energy particle must follow Eq. (3.16).

Consider the spherical coordinates in Fig. 3.8 for each energy particle on a wall element. The ejecting directions (θ, η) of each particle can be determined as follows. For θ, the relation

$$R_\theta = \theta/2\pi \qquad (3.28)$$

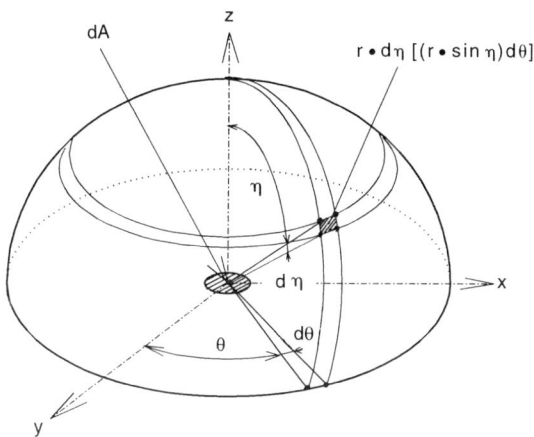

FIG. 3.8. Angles of radiative energy emission from a wall element

holds in a similar manner as Eq. (3.19). In the case of η, if the wall surface is gray and emits a radiative energy of intensity I, the energy emitted from an infinitesimal area dA within an infinitesimal solid angle $d\Omega$ in the direction of the azimuthal angle η can be expressed as $I dA \cos \eta \, d\Omega$, according to Eq. (1.7). It follows from Fig. 3.8 that

$$d\Omega = r \, d\eta (r \sin \eta) \, d\theta / r^2 = \sin \eta \, d\theta \, d\eta. \qquad (3.29)$$

The radiative energy $E_{0,\eta} \, d\eta \, dA$, which is emitted from dA with the azimuthal angle between η and $(\eta + d\eta)$, can be expressed as

$$E_{0,\eta} \, d\eta \, dA = \int_0^{2\pi} d\theta I \, dA \cos \eta \sin \eta \, d\eta$$

$$= 2\pi I \, dA \cos \eta \sin \eta \, d\eta. \qquad (3.30)$$

The total energy that is emitted from dA into the upper hemisphere is obtained from Eq. (1.10) as

$$E \, dA = \pi I \, dA. \qquad (3.31)$$

If the probability density function of η is $f(\eta)$, the probability for the ejecting angle of radiative energy to be between η and $(\eta + d\eta)$ is

$$f(\eta) = E_{0,\eta} \, d\eta \, dA / (E \, dA)$$

$$= 2 \cos \eta \sin \eta \, d\eta. \qquad (3.32)$$

An application of the inverse transformation method produces

$$\xi = \int_0^{\eta} 2 \cos \eta \sin \eta \, d\eta = 1 - \cos^2 \eta. \qquad (3.33)$$

This is the probability for the azimuthal angle of radiative energy emitted from dA to be between 0 and η. Because it is a uniform probability taking a value between 0 and 1, ξ can be replaced by the uniform random number R_η. From Eqs. (3.28) and (3.33), the direction of each energy particle ejected from the wall element can be found by

$$\theta = 2\pi R_\theta, \qquad (3.34)$$

$$\eta = \cos^{-1} \sqrt{1 - R_\eta}. \qquad (3.35)$$

The two uniform random numbers (R_θ, R_η) for each particle are determined using Eqs. (3.34) and (3.35). The ejecting direction distribution of all energy particles ejected from the wall element satisfies Lambert's cosine law, Eq. (1.7).

This section can be summarized as follows: To simulate the behavior of radiative energy emitted from a wall element by means of the Monte Carlo method, a large number of energy particles are ejected from any point on the element surface. Each particle is ejected from the point expressed by Eqs. (3.25)–(3.27) in the direction described by Eqs. (3.34) and (3.35); travels a flight distance described by Eq. (3.16); and is then absorbed by the gas element at that location. From the distribution of all absorption points in the system, the absorption distribution of all radiative energy emitted from the wall element can be determined. Five uniform random numbers per particle are utilized in this analysis.

3.3.4. Simulation of Reflection and Absorption by Solid Walls

Energy particles ejected from a gas element or a wall element in the direction described by Eqs. (3.23) and (3.24) or Eqs. (3.34) and (3.35) travel a distance described by Eq. (3.16) where the energy they possess is completely absorbed by the gas element. It is necessary to consider the reflection and absorption of energy particles by the wall, in the event these particles impact with the wall prior to the completion of the flight distance. Of the radiative energy that is incident on the wall, a fraction corresponding to the wall absorptivity α is absorbed by the wall, and the remaining $(1 - \alpha)$ is reflected. For a gray wall, its absorptivity α is equal to the wall emissivity ε, according to Eq. (1.21). For the sake of simulating this phenomenon by means of the Monte Carlo method, the uniform random number R_r with a magnitude of 0 to 1 is used for each energy particle, to deal with many particles incident on the gray wall of emissivity ε. The particle is absorbed by the wall element when

$$R_r \leq \varepsilon \tag{3.36}$$

and is reflected when

$$R_r > \varepsilon. \tag{3.37}$$

By setting up these conditions, the fraction of an incident energy ($\varepsilon = \alpha$) is absorbed, while the fraction $(1 - \varepsilon) = (1 - \alpha)$ is reflected. The reflection–absorption characteristics of the radiative energy on the wall are thus satisfied. In case the reflection from a gray wall is diffuse, the reflecting direction of each individual particle must be such that those of many energy particles statistically satisfy Lambert's cosine law. This condition can be fulfilled through the determination of the reflecting direction of each energy particle (θ, η) by substituting two uniform random numbers into Eqs. (3.34) and (3.35). Here, the spherical coordinates (θ, η), shown in Fig. 3.8, are fixed on the wall element. If the wall surface is perfectly reflecting, like a mirror surface, the reflecting angle can be geometrically evaluated from the incident angle. The flight distance of a particle after reflection is equal to the total flight distance minus that prior to the reflection.

3.3.5. Radiative Heat Transfer Simulation

The absorption distribution of radiative energy emitted from the gas and wall elements in a system can be determined using the equations presented in Sections 3.3.1 through 3.3.4. This distribution expresses the radiative energy that is transferred from the emitted elements to other elements. Using these results, Eq. (3.5) can provide the radiative heat input term $q_{r,\text{in}}$, which appears in the heat equations (3.1) and (3.2) for the gas and wall elements, respectively.

Figure 3.9 shows the two-dimensional space enclosed by the gray inner walls of a 5- ×5-m square-column-type duct. An example is presented here, using the equations mentioned previously to analyze radiative heat exchanges among the elements by means of the Monte Carlo method. The gas space is divided into $5 \times 5 = 25$ square gas elements, and each wall is divided into 5 elements with a total of 20 wall elements. Although the element division is two dimensional, the emission of radiation energy from both the wall surfaces and the gas is three dimensional, with one component perpendicular to the paper. Hence, the following calculations for the loci of the energy particles are performed as three-dimensional.

3.3.5.1. Uniform Gas Absorption Coefficient Case

Consider the radiative energy emitted from the gas element at the center of Figure 3.10. A fixed number NRAY of energy particles are

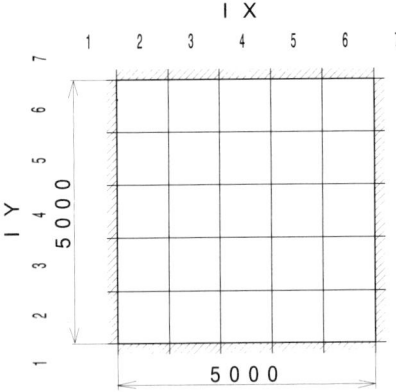

Fig. 3.9. Mesh divisions in a 2-D radiation heat transfer analysis

3.3. SIMULATION OF RADIATIVE HEAT TRANSFER

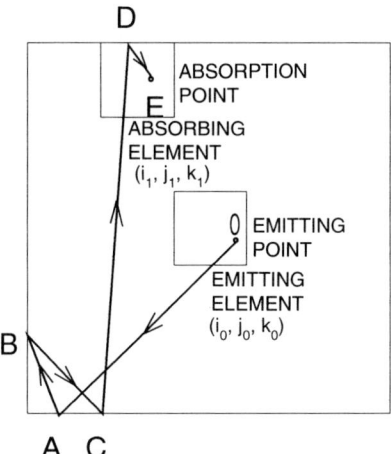

FIG. 3.10. Locus of an energy particle emitted from and absorbed by gas elements ($K = 0.1$ m^{-1}, $\varepsilon = 0.1$)

ejected from this element. Two uniform random numbers are substituted into Eqs. (3.25) and (3.26) to determine the location of the ejection site O, with the ejecting direction and the flight distance determined by Eqs. (3.23) and (3.24) and Eq. (3.16), respectively. Figure 3.10 illustrates an example of the locus of one energy particle being ejected from the central gas element until its absorption. The actual locus is three dimensional with a component perpendicular to the paper. The succeeding figures show two-dimensional loci being projected on the paper surface. The three-dimensional length of (OA + AB + BC + CD + DE) in Fig. 3.10 is the flight length of a particle calculated by Eq. (3.16). If this particle strikes the wall surface before the completion of the flight distance, Eqs. (3.36) and (3.37) are used to decide whether it is absorbed or reflected by the wall element. In Fig. 3.10, the energy particle strikes the wall surface at points A, B, C, and D and is reflected due to a relatively small value of the wall emissivity at $\varepsilon = 0.1$. Due to the assumption of the gray wall in Fig. 3.10, the reflection is diffuse in the direction described by Eqs. (3.34) and (3.35). Hence, the incident and reflection angles at points A, B, C, and D are all different. This energy particle is eventually absorbed by point E inside the gas element, which is located at the upper left portion of the system.

Next we treat the wall in Figure 3.10 as a blackbody, equivalent to $\varepsilon = 1$ in Eq. (3.36). Figure 3.11 shows the absorption of the energy particle by

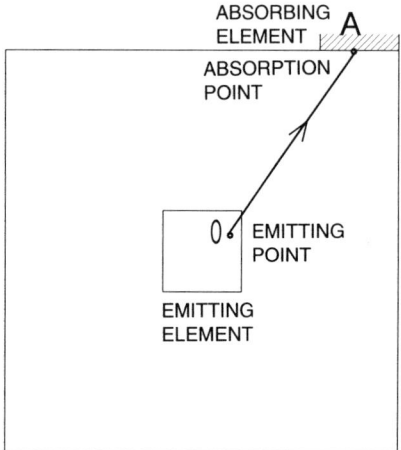

FIG. 3.11. Locus of an energy particle emitted from and absorbed by gas elements ($K = 0.1$ m^{-1}, $\varepsilon = 1.0$)

the wall element upon impact, that is, without reflection. Hence, the locus of the particle has an absorption point on the wall element.

The gas absorption coefficient K takes a low value of 0.1 m^{-1} in Fig. 3.10. As K is increased to 0.7 m^{-1}, the mean flight distance of an energy particle starting from the central gas element is substantially shortened, as shown in Fig. 3.12. Figure 3.13 shows the absorption locations of 10,000 particles ejected from the central gas element. It is seen that due to a high gas absorption coefficient, a substantially large portion of radiative energy (i.e., energy particles) emitted from the central element is absorbed by the element itself or in its vicinity.

Another example treats the locus of an energy particle ejected from the wall element at the center of the left wall. Under the conditions of $K = 0.7$ m^{-1} and $\varepsilon = 0.1$, Fig. 3.14 illustrates the locus of an energy particle. Figure 3.15 illustrates the distribution of absorption locations of 10,000 particles for the conditions of $K = 0.7$ m^{-1} and $\varepsilon = 1$.

These figures are obtained from a computer program written in PC BASIC, which differs greatly depending on the type of PC used. Figure 3.16 is a computer program written in FORTRAN 77, which is common worldwide. The variables in the program, as well as others used in the text, are presented at the end of this monograph under the title List of Variables in Computer Programs. The program RAT1, shown in Figure 3.16, calculates the absorption distribution of energy particles ejected from an arbitrarily selected gas element or wall element inside the system of

3.3. SIMULATION OF RADIATIVE HEAT TRANSFER

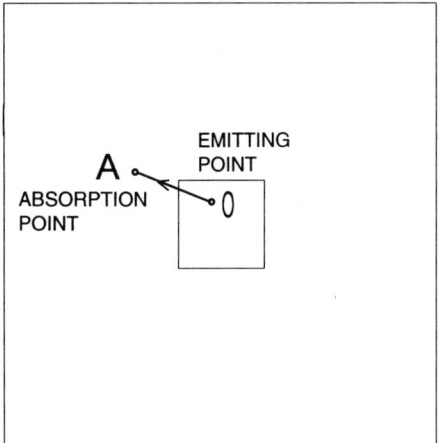

FIG. 3.12. Locus of an energy particle emitted from and absorbed by gas elements ($K = 0.8$ m^{-1}, $\varepsilon = 1.0$)

Fig. 3.9. Line 20 reads in the gas absorption coefficient AK (unit of inverse meters), wall emissivity EM, number of energy particles NRAY, and identification of ejecting element (IX, IY). One may refer to Fig. 3.9 for IX, IY = 1 through 7. Lines 53 through 239 identify the element where each energy particle from the ejecting element is eventually absorbed. The

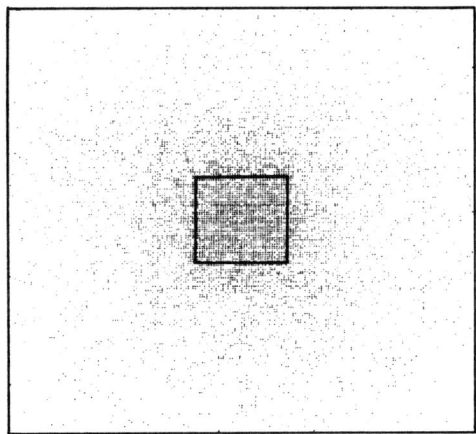

FIG. 3.13. Distribution of absorption locations of radiative energy particles ($K = 0.7$ m^{-1}, $\varepsilon = 1.0$)

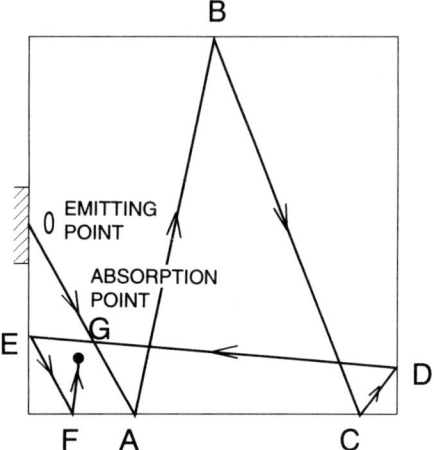

FIG. 3.14. Locus of an energy particle emitted from a wall and absorbed by gas elements ($K = 0.1 \text{ m}^{-1}$, $\varepsilon = 0.1$)

procedure begins with line 47, the DO loop, repeating NRAY times (number of ejected energy particles) to determine the distribution of the absorption locations of NRAY particles. Figure 3.17 is the flowchart for the principal part of RAT1 (lines 53 through 239) to determine the absorption location of one energy particle. The number at the upper left

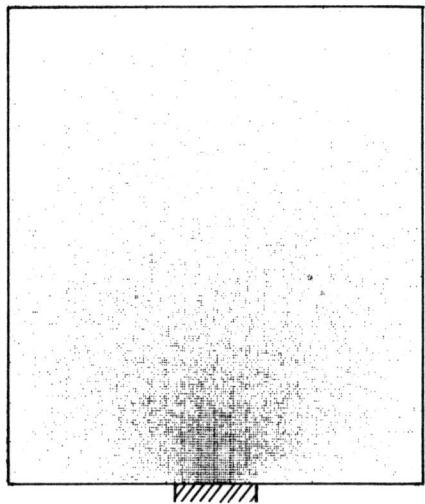

FIG. 3.15. Distribution of absorption locations of radiative energy particles ($K = 0.7 \text{ m}^{-1}$, $\varepsilon = 1.0$)

3.3. SIMULATION OF RADIATIVE HEAT TRANSFER

```
1     ***********************************************************************
2     *                                                                     *
3     *                              RAT1                                   *
4     *                                                                     *
5     *           MONTE CARLO SIMULATION OF RADIATIVE TRANSFER              *
6     *                (UNIFORM GAS ABSORPTION COEFFICIENT)                 *
7     ***********************************************************************
8     C
9           DIMENSION NRD(7,7),RD(7,7)
10          REAL*8 RAND
11          DATA NRD/49*0/,RD/49*0./
12          DATA DXG,DYG/1.0,1.0/
13          RAND=5249347.0D0
14          open ( 6,file='PRN' )
15          write(*,100)
16      100 format (1h0, 'input absorption coeff. of gas (1/m)'
17         *          /'           wall emissivity'
18         *          /'           number of energy particles'
19         *          /'           position of emitting element (1-7,1-7)')
20          READ(*,*) AK,EM,NRAY,IX,IY
21          PAI=3.14159
22          IF ((IX.EQ.1).OR.(IX.EQ.7).OR.(IY.EQ.1).OR.(IY.EQ.7)) THEN
23             KA=1
24          ELSE
25             KA=0
26          ENDIF
27          XC=(FLOAT(IX-1)-0.5)*DXG
28          YC=(FLOAT(IY-1)-0.5)*DYG
29          WRITE(6,200)
30      200 FORMAT(1H0,'*** MONTE CARLO SIMULATION OF RADIATIVE TRANSFER ***
31         *'/)
32          WRITE(6,210) AK
33      210 FORMAT(1H ,'   GAS ABSORPTION COEFF. (1/M)',E13.5)
34          WRITE(6,220) EM
35      220 FORMAT(1H ,'   WALL EMISSIVITY              ',E13.5)
36          WRITE(6,230) NRAY
37      230 FORMAT(1H ,'   NUMBER OF ENERGY PARTICLES   ',I7)
38          WRITE(6,240) IX,IY
39      240 FORMAT(1H ,'   EMITTING ELEMENT            (',I1,',',I1,')')
40          WRITE(6,245) DXG,DYG
41      245 FORMAT(1H ,'   ELEMENT SIZE                  DXG=',F9.5,'(m)'
42         *          /'                                 DYG=',F9.5,'(m)'//)
43    ***********************************************************************
44    *       PURSUIT OF ENERGY PARTICLES
45    ***********************************************************************
46          ndisp=0
47          DO 1000 INRAY=1,NRAY
48          ndisp=ndisp+1
49          if(ndisp.eq.1000) then
50             write(*,*) inray
51             ndisp=0
52          endif
53    *----------------------------------------------------------------------
54    *       DECISION OF EMITTING POINTS
55    *----------------------------------------------------------------------
56          INDABS=0
57          CALL RANDOM(RAN,RAND)
58          S=-ALOG(1.0-RAN)/AK
59          CALL RANDOM(RAN,RAND)
```

FIG. 3.16. Program of radiative transfer simulation by Monte Carlo method (uniform property value case)

```
 60              THTA=2.0*PAI*RAN
 61              IF (KA.EQ.1) THEN
 62     *        ------------------------
 63     *        (EMISSION FROM WALL)
 64     *        ------------------------
 65                CALL RANDOM(RAN,RAND)
 66                ETAW=ACOS(SQRT(1.0-RAN))
 67                CALL RANDOM(RAN,RAND)
 68                IF(IX.EQ.1) THEN
 69                  X0=0.0
 70                  Y0=(RAN-0.5)*DYG+YC
 71                  XE=X0+S*COS(ETAW)
 72                  YE=Y0+S*SIN(ETAW)*SIN(THTA)
 73                ELSEIF(IX.EQ.7) THEN
 74                  X0=5.0*DXG
 75                  Y0=(RAN-0.5)*DYG+YC
 76                  XE=X0-S*COS(ETAW)
 77                  YE=Y0+S*SIN(ETAW)*SIN(THTA)
 78                ELSEIF(IY.EQ.1) THEN
 79                  X0=(RAN-0.5)*DXG+XC
 80                  Y0=0.0
 81                  XE=X0+S*SIN(ETAW)*COS(THTA)
 82                  YE=Y0+S*COS(ETAW)
 83                ELSEIF(IY.EQ.7) THEN
 84                  X0=(RAN-0.5)*DXG+XC
 85                  Y0=5.0*DYG
 86                  XE=X0+S*SIN(ETAW)*COS(THTA)
 87                  YE=Y0-S*COS(ETAW)
 88                ENDIF
 89     *        ----------------------
 90     *        (EMISSION FROM GAS)
 91     *        ----------------------
 92              ELSE
 93                CALL RANDOM(RAN,RAND)
 94                ETAG=ACOS(1.0-2.0*RAN)
 95                CALL RANDOM(RAN,RAND)
 96                X0=(RAN-0.5)*DXG+XC
 97                CALL RANDOM(RAN,RAND)
 98                Y0=(RAN-0.5)*DYG+YC
 99                XE=X0+S*SIN(ETAG)*COS(THTA)
100                YE=Y0+S*COS(ETAG)
101              ENDIF
102     *----------------------------------------------------------------
103     *        DECISION OF ABSORPTION POINT
104     *----------------------------------------------------------------
105     5000     CONTINUE
106     *        ----------------------
107     *        (HIT ON THE WALLS)
108     *        ----------------------
109              IF(YE.LT.0.0) THEN
110                YW=0.0
111                XW=X0+(YW-Y0)*(XE-X0)/(YE-Y0)
112                IF(XW.LT.0.0) THEN
113                  XW=0.0
114                  YW=Y0+(XW-X0)*(YE-Y0)/(XE-X0)
115                CALL RANDOM(RAN,RAND)
116                IF(RAN.LT.EM) THEN
117                  X1=XW
118                  Y1=YW
```

FIG. 3.16. (*Continued*)

3.3. SIMULATION OF RADIATIVE HEAT TRANSFER

```
119             INDABS=1
120           ENDIF
121         ELSEIF(XW.GT.5.0*DXG) THEN
122           XW=5.0*DXG
123           YW=Y0+(XW-X0)*(YE-Y0)/(XE-X0)
124           CALL RANDOM(RAN,RAND)
125           IF(RAN.LT.EM) THEN
126             X1=XW
127             Y1=YW
128             INDABS=3
129           ENDIF
130         ELSE
131           CALL RANDOM(RAN,RAND)
132           IF(RAN.LT.EM) THEN
133             X1=XW
134             Y1=YW
135             INDABS=4
136           ENDIF
137         ENDIF
138       ELSEIF(YE.GT.5.0*DYG) THEN
139         YW=5.0*DYG
140         XW=X0+(YW-Y0)*(XE-X0)/(YE-Y0)
141         IF(XW.LT.0.0) THEN
142           XW=0.0
143           YW=Y0+(XW-X0)*(YE-Y0)/(XE-X0)
144           CALL RANDOM(RAN,RAND)
145           IF(RAN.LT.EM) THEN
146             X1=XW
147             Y1=YW
148             INDABS=1
149           ENDIF
150         ELSEIF(XW.GT.5.0*DXG) THEN
151           XW=5.0*DXG
152           YW=Y0+(XW-X0)*(YE-Y0)/(XE-X0)
153           CALL RANDOM(RAN,RAND)
154           IF(RAN.LT.EM) THEN
155             X1=XW
156             Y1=YW
157             INDABS=3
158           ENDIF
159         ELSE
160           CALL RANDOM(RAN,RAND)
161           IF(RAN.LT.EM) THEN
162             X1=XW
163             Y1=YW
164             INDABS=2
165           ENDIF
166         ENDIF
167       ELSEIF(XE.LT.0.0) THEN
168         XW=0.0
169         YW=Y0+(XW-X0)*(YE-Y0)/(XE-X0)
170         CALL RANDOM(RAN,RAND)
171         IF(RAN.LT.EM) THEN
172           X1=XW
173           Y1=YW
174           INDABS=1
175         ENDIF
176       ELSEIF(XE.GT.5.0*DXG) THEN
177         XW=5.0*DXG
```

FIG. 3.16. (*Continued*)

```
178            YW=Y0+(XW-X0)*(YE-Y0)/(XE-X0)
179            CALL RANDOM(RAN,RAND)
180            IF(RAN.LT.EM) THEN
181              X1=XW
182              Y1=YW
183              INDABS=3
184            ENDIF
185  *     ----------------------
186  *        (ABSORPTION BY GAS)
187  *     ----------------------
188          ELSE
189            X1=XE
190            Y1=YE
191            INDABS=5
192          ENDIF
193          IF(INDABS.NE.0) GOTO 5010
194  *----------------------------------------------------------------
195  *        DECISION OF REFLECTION DIRECTION AT WALL
196  *----------------------------------------------------------------
197          S=S-SQRT((XW-X0)**2+(YW-Y0)**2)
198          X0=XW
199          Y0=YW
200          CALL RANDOM(RAN,RAND)
201          THTA=2.0*PAI*RAN
202          CALL RANDOM(RAN,RAND)
203          ETAW=ACOS(SQRT(1.0-RAN))
204          IF(YW.EQ.0.0) THEN
205            XE=X0+S*SIN(ETAW)*SIN(THTA)
206            YE=Y0+S*COS(ETAW)
207          ELSEIF(YW.EQ.5.0*DYG) THEN
208            XE=X0+S*SIN(ETAW)*SIN(THTA)
209            YE=Y0-S*COS(ETAW)
210          ELSEIF(XW.EQ.0.0) THEN
211            XE=X0+S*COS(ETAW)
212            YE=Y0+S*SIN(ETAW)*SIN(THTA)
213          ELSE
214            XE=X0-S*COS(ETAW)
215            YE=Y0+S*SIN(ETAW)*SIN(THTA)
216          ENDIF
217          GOTO 5000
218  ****************************************************************
219  *        COUNTING THE ABSORBED PARTICLE NUMBERS
220  ****************************************************************
221  5010    CONTINUE
222          IF(INDABS.EQ.1) THEN
223            IXA=1
224            IYA=IFIX(Y1/DYG)+2
225          ELSEIF(INDABS.EQ.2) THEN
226            IXA=IFIX(X1/DXG)+2
227            IYA=7
228          ELSEIF(INDABS.EQ.3) THEN
229            IXA=7
230            IYA=IFIX(Y1/DYG)+2
231          ELSEIF(INDABS.EQ.4) THEN
232            IXA=IFIX(X1/DXG)+2
233            IYA=1
234          ELSE
235            IXA=IFIX(X1/DXG)+2
236            IYA=IFIX(Y1/DYG)+2
```

FIG. 3.16. (*Continued*)

3.3. SIMULATION OF RADIATIVE HEAT TRANSFER

```
237            ENDIF
238            NRD(IXA,IYA)=NRD(IXA,IYA)+1
239  1000 CONTINUE
240  ************************************************************
241  *       PRINT RESULTS
242  ************************************************************
243            ANRAY=FLOAT(NRAY)
244            ASN=FLOAT(NRD(IX,IY))
245            AS=ASN/ANRAY
246            OUTRAY=ANRAY-ASN
247            DO 1010 I=1,7
248              DO 1020 J=1,7
249                IF((I.EQ.IX).AND.(J.EQ.IY)) THEN
250                  RD(I,J)=AS
251                ELSE
252                  RD(I,J)=FLOAT(NRD(I,J))/OUTRAY
253                ENDIF
254  1020      CONTINUE
255  1010    CONTINUE
256   250 FORMAT(1H ,11X,5(I7,4X))
257   260 FORMAT(1H ,7(I7,4X))
258   270 FORMAT(1H ,11X,5(F9.5,2X))
259   280 FORMAT(1H ,7(F9.5,2X))
260       WRITE(6,290)
261   290 FORMAT(1H0,5X,'(NUMBER OF ABSORBED ENERGY PARTICLES)'/)
262       WRITE(6,250) NRD(2,7),NRD(3,7),NRD(4,7),NRD(5,7),NRD(6,7)
263       DO 1030 I=6,2,-1
264         WRITE(6,260) NRD(1,I),NRD(2,I),NRD(3,I),NRD(4,I),NRD(5,I),
265      *               NRD(6,I),NRD(7,I)
266  1030 CONTINUE
267       WRITE(6,250) NRD(2,1),NRD(3,1),NRD(4,1),NRD(5,1),NRD(6,1)
268       WRITE(6,300)
269   300 FORMAT(1H0,5X,'(RELATIVE ABSORBED ENERGY PROFILE)'/)
270       WRITE(6,270) RD(2,7),RD(3,7),RD(4,7),RD(5,7),RD(6,7)
271       DO 1040 I=6,2,-1
272         WRITE(6,280) RD(1,I),RD(2,I),RD(3,I),RD(4,I),RD(5,I),
273      *               RD(6,I),RD(7,I)
274  1040 CONTINUE
275       WRITE(6,270) RD(2,1),RD(3,1),RD(4,1),RD(5,1),RD(6,1)
276       STOP
277       END
278  *
279  ************************************************************
280  *       RANDOM NUMBER GENERATOR
281  ************************************************************
282       SUBROUTINE RANDOM(RAN,RAND)
283       REAL*8 RAND
284       RAND=DMOD(RAND*131075.0D0,2147483649.0D0)
285       RAN=SNGL(RAND/2147483649.0D0)
286       RETURN
287       END
```

FIG. 3.16. (*Continued*)

corner outside each block in Fig. 3.17 corresponds to the line number in the program of Fig. 3.16.

An example of the output of the program is displayed in Fig. 3.18, the results for 100,000 energy particles being ejected from the central gas element (4,4) in the system with $K = 0.1$ m^{-1} and $\varepsilon = 0.1$. The upper half

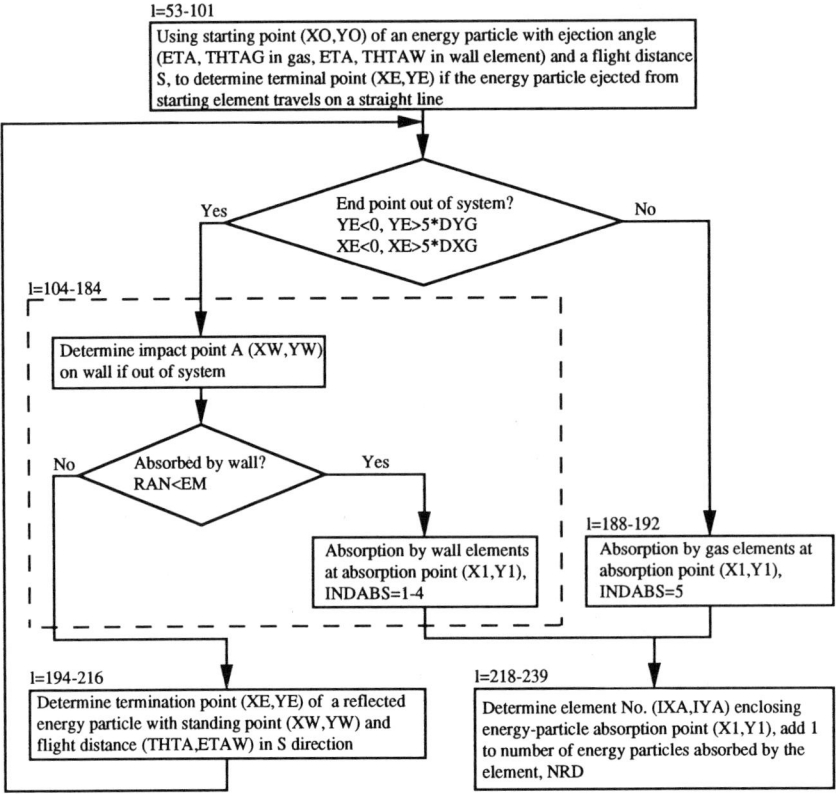

FIG. 3.17. Flowchart of principal part of program RAT1 (lines 53 through 239) of Fig. 3.16

presents the number of energy particles absorbed by each element, which is arranged in the order of Fig. 3.9. The lower half lists the magnitude of $\alpha_{s,g}(4,4)$, and $R_d(4,4,i,j)$, representing the fraction of self-absorption, or energy absorbed by other elements, of the radiative energy emitted from the gas element (4,4). Its corresponding READ (R_d) numbers are given in Fig. 3.19 and $\alpha_{s,g}$ and R_d are used in Eqs. (3.3) and (3.5), respectively. Their magnitudes are determined by Eqs. (4.3) and (4.4), respectively, as explained in Chapter 4. Noted that in Figs. 3.18 and 3.19 the gas elements are within the larger rectangle, with the wall elements outside. The small rectangle corresponds to the central gas element from which radiative energy is emitted. The $\alpha_{s,g}(4,4)$ in the center of Fig. 3.19 signifies the

3.3. SIMULATION OF RADIATIVE HEAT TRANSFER

```
*** MONTE CARLO SIMULATION OF RADIATIVE TRANSFER ***

GAS ABSORPTION COEFF. (1/M)      .10000E+00
WALL EMISSIVITY                  .10000E+00
NUMBER OF ENERGY PARTICLES       100000
EMITTING ELEMENT                 (4,4)
ELEMENT SIZE                     DXG=   1.00000(m)
                                 DYG=   1.00000(m)
```

(NUMBER OF ABSORBED ENERGY PARTICLES)

	796	883	937	896	822	
827	2340	2791	2870	2781	2306	758
877	2589	3564	4562	3547	2646	900
946	2833	4327	8992	4311	2721	913
847	2743	3568	4361	3485	2699	908
822	2250	2624	2750	2700	2254	774
	795	925	990	948	822	

(RELATIVE ABSORBED ENERGY PROFILE)

	.00875	.00970	.01030	.00985	.00903	
.00909	.02571	.03067	.03154	.03056	.02534	.00833
.00964	.02845	.03916	.05013	.03897	.02907	.00989
.01039	.03113	.04755	.08992	.04737	.02990	.01003
.00931	.03014	.03921	.04792	.03829	.02966	.00998
.00903	.02472	.02883	.03022	.02967	.02477	.00850
	.00874	.01016	.01088	.01042	.00903	

FIG. 3.18. Output of RAT1 program, example 1

fraction of the radiative energy emitted from the gas element (4,4), which is absorbed by the element itself. Its magnitude is 0.08992 in the example of Fig. 3.18. The R_d (4,4,3,3) entry located immediately above the central element in Fig. 3.19 gives the fraction of the radiative energy emitted outside from the gas element (4,4) absorbed by the element (4,3). The

	Rd(4,4,2,1)	Rd(4,4,3,1)	Rd(4,4,4,1)	Rd(4,4,5,1)	Rd(4,4,6,1)	
Rd(4,4,1,2)	Rd(4,4,2,2)	Rd(4,4,3,2)	Rd(4,4,4,2)	Rd(4,4,5,2)	Rd(4,4,6,2)	Rd(4,4,7,2)
Rd(4,4,1,3)	Rd(4,4,2,3)	Rd(4,4,3,3)	Rd(4,4,4,3)	Rd(4,4,5,3)	Rd(4,4,6,3)	Rd(4,4,7,3)
Rd(4,4,1,4)	Rd(4,4,2,4)	Rd(4,4,3,4)	$\alpha_{s,g}$ (4,4)	Rd(4,4,5,4)	Rd(4,4,6,4)	Rd(4,4,7,4)
Rd(4,4,1,5)	Rd(4,4,2,5)	Rd(4,4,3,5)	Rd(4,4,4,5)	Rd(4,4,5,5)	Rd(4,4,6,5)	Rd(4,4,7,5)
Rd(4,4,1,6)	Rd(4,4,2,6)	Rd(4,4,3,6)	Rd(4,4,4,6)	Rd(4,4,5,6)	Rd(4,4,6,6)	Rd(4,4,7,6)
	Rd(4,4,2,7)	Rd(4,4,3,7)	Rd(4,4,4,7)	Rd(4,4,5,7)	Rd(4,4,6,7)	

FIG. 3.19. READ numbers

```
*** MONTE CARLO SIMULATION OF RADIATIVE TRANSFER ***
GAS ABSORPTION COEFF. (1/M)    .70000E+00
WALL EMISSIVITY                .10000E+01
NUMBER OF ENERGY PARTICLES      100000
EMITTING ELEMENT               (4,4)
ELEMENT SIZE                   DXG=  1.00000(m)
                               DYG=  1.00000(m)


      (NUMBER OF ABSORBED ENERGY PARTICLES)
                234       567       689       615       278
     249        431       910      1243       908       424        247
     540        905      3241      7314      3247       949        578
     759       1246      7321     34202      7347      1265        698
     549        859      3111      7396      3314       923        527
     271        449       950      1258       896       424        271
                264       542       746       576       267

      (RELATIVE ABSORBED ENERGY PROFILE)
              .00356    .00862    .01047    .00935    .00423
   .00378    .00655    .01383    .01889    .01380    .00644     .00375
   .00821    .01375    .04926    .11116    .04935    .01442     .00878
   .01154    .01894    .11126    .34202    .11166    .01923     .01061
   .00834    .01306    .04728    .11240    .05037    .01403     .00801
   .00412    .00682    .01444    .01912    .01362    .00644     .00412
              .00401    .00824    .01134    .00875    .00406
```

FIG. 3.20. Output of RAT1 program, example 2

value of R_d is 0.05013 in the example of Fig. 3.18. Because of a low gas absorption coefficient, $K = 0.1$ m^{-1}, the radiative energy attenuates slowly and consequently reaches a long distance. In addition, a low wall emissivity, $\varepsilon = 0.1$, causes a possible occurrence of multiple reflections. Hence, the radiative energy emitted from the central gas element is relatively, uniformly absorbed within the system.

In contrast, when the gas absorption coefficient is raised to 0.7 m^{-1} and the wall emissivity to unity, Fig. 3.20 indicates that most of the radiative energy emitted from the central gas element will be absorbed by the emitting gas element (4,4) itself and its vicinity. Since the analysis is conducted under the same conditions as in Fig. 3.13, the analytical results in Fig. 3.20 corresponds to the absorption distribution in Fig. 3.13. In this case, the fraction of self-absorption by the emitting gas element is 0.34202, or 3.8 times the example of Fig. 3.18. This means the emitting element itself captures as much as 34% of the radiative energy it has released.

Figure 3.21 presents the analytical results for the absorption distribution of radiative energy emitted from the wall element (1,4) under the same conditions as Fig. 3.20. Because the analysis is conducted under the same

3.3. SIMULATION OF RADIATIVE HEAT TRANSFER

```
*** MONTE CARLO SIMULATION OF RADIATIVE TRANSFER ***

GAS ABSORPTION COEFF. (1/M)    .70000E+00
WALL EMISSIVITY                .10000E+01
NUMBER OF ENERGY PARTICLES     100000
EMITTING ELEMENT               (1,4)
ELEMENT SIZE                   DXG=   1.00000(m)
                               DYG=   1.00000(m)
```

(NUMBER OF ABSORBED ENERGY PARTICLES)

		488	689	418	212	115	
0		1040	1360	747	302	114	108
0		9012	4353	1514	590	186	114
0		46487	7641	2087	682	238	175
0		8915	4397	1491	542	212	149
0		1015	1336	793	294	136	118
		505	689	418	214	104	

(RELATIVE ABSORBED ENERGY PROFILE)

		.00488	.00689	.00418	.00212	.00115	
.00000		.01040	.01360	.00747	.00302	.00114	.00108
.00000		.09012	.04353	.01514	.00590	.00186	.00114
.00000		.46487	.07641	.02087	.00682	.00238	.00175
.00000		.08915	.04397	.01491	.00542	.00212	.00149
.00000		.01015	.01336	.00793	.00294	.00136	.00118
		.00505	.00689	.00418	.00214	.00104	

FIG. 3.21. Output of RAT1 program, example 3

```
*** MONTE CARLO SIMULATION OF RADIATIVE TRANSFER ***

GAS ABSORPTION COEFF. (1/M)    .70000E+00
WALL EMISSIVITY                .10000E+01
NUMBER OF ENERGY PARTICLES     100000
EMITTING ELEMENT               (4,4)
ELEMENT SIZE                   DXG=   5.00000(m)
                               DYG=   5.00000(m)
```

(NUMBER OF ABSORBED ENERGY PARTICLES)

		0	0	0	0	0	
0		0	19	43	11	0	0
0		15	802	5274	778	13	0
0		52	5324	75365	5261	38	0
0		14	791	5328	789	4	0
0		2	22	36	18	0	0
		0	1	0	0	0	

(RELATIVE ABSORBED ENERGY PROFILE)

		.00000	.00000	.00000	.00000	.00000	
.00000		.00000	.00077	.00175	.00045	.00000	.00000
.00000		.00061	.03256	.21409	.03158	.00053	.00000
.00000		.00211	.21612	.75365	.21356	.00154	.00000
.00000		.00057	.03211	.21628	.03203	.00016	.00000
.00000		.00008	.00089	.00146	.00073	.00000	.00000
		.00000	.00004	.00000	.00000	.00000	

FIG. 3.22. Output of RAT1 program, example 4

```
  1  ************************************************************************
  2  *                                                                      *
  3  *                              RAT2                                    *
  4  *                                                                      *
  5  *         MONTE CARLO SIMULATION OF RADIATIVE TRANSFER                 *
  6  *            (DISTRIBUTED GAS ABSORPTION COEFFICIENT)                  *
  7  ************************************************************************
  8  C
  9        DIMENSION NRD(7,7),RD(7,7),AK(7,7),S(4)
 10        REAL*8 RAND
 11        DATA NRD/49*0/,RD/49*0./
 12        DATA AK/7*-1.0,
 13       *            -1.0,  2*0.1,    0.7,  2*0.1, -1.0,
 14       *            -1.0,    0.1,  3*0.7,    0.1, -1.0,
 15       *            -1.0,    0.1,  3*0.7,    0.1, -1.0,
 16       *            -1.0,           5*0.1,        -1.0,
 17       *            -1.0,           5*0.1,        -1.0,
 18       *          7*-1.0/
 19        DATA DXG,DYG/1.0,1.0/
 20        RAND=5249347.0D0
 21        open ( 6,file='PRN' )
 22        write(*,100)
 23    100 format (1h0, 'input wall emissivity'
 24       *          /'         number of energy particles'
 25       *          /'         position of emitting element (1-7,1-7)')
 26        READ(*,*) EM,NRAY,IX,IY
 27        PAI=3.14159
 28        IF ((IX.EQ.1).OR.(IX.EQ.7).OR.(IY.EQ.1).OR.(IY.EQ.7)) THEN
 29           KA=1
 30        ELSE
 31           KA=0
 32        ENDIF
 33        XC=(FLOAT(IX-1)-0.5)*DXG
 34        YC=(FLOAT(IY-1)-0.5)*DYG
 35        WRITE(6,200)
 36    200 FORMAT(1H0,'*** MONTE CARLO SIMULATION OF RADIATIVE TRANSFER ***
 37       *'/)
 38        WRITE(6,210)
 39    210 FORMAT(1H ,'   GAS ABSORPTION COEFF. (1/m)')
 40        DO 900 I=7,1,-1
 41           WRITE(6,215) AK(1,I),AK(2,I),AK(3,I),AK(4,I),AK(5,I),
 42       *                AK(6,I),AK(7,I)
 43    215    FORMAT(1H ,5X,7(F9.5,2X))
 44    900 CONTINUE
 45        WRITE(6,220) EM
 46    220 FORMAT(1H ,'   WALL EMISSIVITY              ',E13.5)
 47        WRITE(6,230) NRAY
 48    230 FORMAT(1H ,'   NUMBER OF ENERGY PARTICLES   ',I7)
 49        WRITE(6,240) IX,IY
 50    240 FORMAT(1H ,'   EMITTING ELEMENT             (',I1,',',I1,')')
 51        WRITE(6,245) DXG,DYG
 52    245 FORMAT(1H ,'   ELEMENT SIZE                 DXG=',F9.5,'(m)'
 53       *          /'                                 DYG=',F9.5,'(m)'//)
 54  ************************************************************************
 55  *         PURSUIT OF ENERGY PARTICLES
 56  ************************************************************************
 57        ndisp=0
 58        DO 1000 INRAY=1,NRAY
 59           ndisp=ndisp+1
```

FIG. 3.23. Program of radiative transfer simulation by Monte Carlo method (case for distributed property values), RAT2

3.3. SIMULATION OF RADIATIVE HEAT TRANSFER

```
60              if(ndisp.eq.1000) then
61                write(*,*) inray
62                ndisp=0
63              endif
64   *------------------------------------------------------------------
65   *      DECISION OF EMITTING POINTS
66   *------------------------------------------------------------------
67              INDABS=0
68              INDGWC=1
69              CALL RANDOM(RAN,RAND)
70              XK=-ALOG(1.0-RAN)
71              CALL RANDOM(RAN,RAND)
72              THTA=2.0*PAI*RAN
73              IF (KA.EQ.1) THEN
74   *          -------------------------
75   *          (EMISSION FROM WALL)
76   *          -------------------------
77                CALL RANDOM(RAN,RAND)
78                ETAW=ACOS(SQRT(1.0-RAN))
79                CALL RANDOM(RAN,RAND)
80                IF(IX.EQ.1) THEN
81                  X0=0.0
82                  Y0=(RAN-0.5)*DYG+YC
83                  AL=COS(ETAW)
84                  AM=SIN(ETAW)*SIN(THTA)
85                  IXT=2
86                  IYT=IY
87                ELSEIF(IX.EQ.7) THEN
88                  X0=5.0*DXG
89                  Y0=(RAN-0.5)*DYG+YC
90                  AL=-COS(ETAW)
91                  AM=SIN(ETAW)*SIN(THTA)
92                  IXT=6
93                  IYT=IY
94                ELSEIF(IY.EQ.1) THEN
95                  X0=(RAN-0.5)*DXG+XC
96                  Y0=0.0
97                  AL=SIN(ETAW)*COS(THTA)
98                  AM=COS(ETAW)
99                  IXT=IX
100                 IYT=2
101               ELSEIF(IY.EQ.7) THEN
102                 X0=(RAN-0.5)*DXG+XC
103                 Y0=5.0*DYG
104                 AL=SIN(ETAW)*COS(THTA)
105                 AM=-COS(ETAW)
106                 IXT=IX
107                 IYT=6
108               ENDIF
109  *          -------------------------
110  *          (EMISSION FROM GAS)
111  *          -------------------------
112             ELSE
113               CALL RANDOM(RAN,RAND)
114               ETAG=ACOS(1.0-2.0*RAN)
115               CALL RANDOM(RAN,RAND)
116               X0=(RAN-0.5)*DXG+XC
117               CALL RANDOM(RAN,RAND)
118               Y0=(RAN-0.5)*DYG+YC
```

FIG. 3.23. (*Continued*)

```
119              AL=SIN(ETAG)*COS(THTA)
120              AM=COS(ETAG)
121              IF(ABS(AL).LT.1.E-10) THEN
122                 AL=SIGN(1.E-10,AL)
123              ENDIF
124              IF(ABS(AM).LT.1.E-10) THEN
125                 AM=SIGN(1.E-10,AM)
126              ENDIF
127              IXT=IX
128              IYT=IY
129           ENDIF
130           XCT=(FLOAT(IXT-1)-0.5)*DXG
131           YCT=(FLOAT(IYT-1)-0.5)*DYG
132           XI=X0-XCT
133           YI=Y0-YCT
134     *-----------------------------------------------------------
135     *     DECISION OF ABSORPTION POINT
136     *-----------------------------------------------------------
137     5000  CONTINUE
138           IF(INDGWC.EQ.1) THEN
139     *        ---------------------------------------------
140     *        (RADIATION TRANSFER THROUGH GAS ELEMENTS)
141     *        ---------------------------------------------
142              S(1)=-(0.5*DXG+XI)/AL
143              S(2)=(0.5*DYG-YI)/AM
144              S(3)=(0.5*DXG-XI)/AL
145              S(4)=-(0.5*DYG+YI)/AM
146              SMIN=1.E20
147              DO 1002 I=1,4
148                 IF((S(I).GT.1.E-4).AND.(S(I).LT.SMIN)) THEN
149                    SMIN=S(I)
150                    IW=I
151                 ENDIF
152     1002     CONTINUE
153              XK=XK-SMIN*AK(IXT,IYT)
154              IF(XK.LE.0.0) THEN
155                 INDABS=5
156                 IXA=IXT
157                 IYA=IYT
158              ELSE
159                 XE=XI+SMIN*AL
160                 YE=YI+SMIN*AM
161                 IF(IW.EQ.1) THEN
162                    IXT=IXT-1
163                 ELSEIF(IW.EQ.2) THEN
164                    IYT=IYT+1
165                 ELSEIF(IW.EQ.3) THEN
166                    IXT=IXT+1
167                 ELSE
168                    IYT=IYT-1
169                 ENDIF
170                 IF((IXT.EQ.1).OR.(IXT.EQ.7).OR.
171     *             (IYT.EQ.1).OR.(IYT.EQ.7)) THEN
172                    INDGWC=0
173                 ELSE
174                    INDGWC=1
175                    IF(IW.EQ.1) THEN
176                       XI=0.5*DXG
177                       YI=YE
```

FIG. 3.23. (*Continued*)

3.3. SIMULATION OF RADIATIVE HEAT TRANSFER

```
178                  ELSEIF(IW.EQ.2) THEN
179                      XI=XE
180                      YI=-0.5*DYG
181                  ELSEIF(IW.EQ.3) THEN
182                      XI=-0.5*DXG
183                      YI=YE
184                  ELSE
185                      XI=XE
186                      YI=0.5*DYG
187                  ENDIF
188                ENDIF
189              ENDIF
190            ELSE
191  *         --------------------
192  *         (HIT ON THE WALLS)
193  *         --------------------
194              CALL RANDOM(RAN,RAND)
195              IF(RAN.LT.EM) THEN
196                INDABS=IW
197                IXA=IXT
198                IYA=IYT
199              ELSE
200                CALL RANDOM(RAN,RAND)
201                ETAW=ACOS(SQRT(1.0-RAN))
202                CALL RANDOM(RAN,RAND)
203                THTA=2.0*PAI*RAN
204                IF(IW.EQ.1) THEN
205                    XI=-0.5*DXG
206                    YI=YE
207                    AL=COS(ETAW)
208                    AM=SIN(ETAW)*SIN(THTA)
209                    IXT=2
210                ELSEIF(IW.EQ.2) THEN
211                    XI=XE
212                    YI=0.5*DYG
213                    AL=SIN(ETAW)*COS(THTA)
214                    AM=-COS(ETAW)
215                    IYT=6
216                ELSEIF(IW.EQ.3) THEN
217                    XI=0.5*DXG
218                    YI=YE
219                    AL=-COS(ETAW)
220                    AM=SIN(ETAW)*SIN(THTA)
221                    IXT=6
222                ELSE
223                    XI=XE
224                    YI=-0.5*DYG
225                    AL=SIN(ETAW)*COS(THTA)
226                    AM=COS(ETAW)
227                    IYT=2
228                ENDIF
229              ENDIF
230            ENDIF
231            IF(INDABS.EQ.0) GOTO 5000
232            NRD(IXA,IYA)=NRD(IXA,IYA)+1
233  1000  CONTINUE
234  *****************************************************************
235  *     PRINT RESULTS
236  *****************************************************************
```

FIG. 3.23. (*Continued*)

```
237         ANRAY=FLOAT(NRAY)
238         ASN=FLOAT(NRD(IX,IY))
239         AS=ASN/ANRAY
240         OUTRAY=ANRAY-ASN
241         DO 1010 I=1,7
242           DO 1020 J=1,7
243             IF((I.EQ.IX).AND.(J.EQ.IY)) THEN
244               RD(I,J)=AS
245             ELSE
246               RD(I,J)=FLOAT(NRD(I,J))/OUTRAY
247             ENDIF
248  1020   CONTINUE
249  1010 CONTINUE
250   250 FORMAT(1H ,11X,5(I7,4X))
251   260 FORMAT(1H ,7(I7,4X))
252   270 FORMAT(1H ,11X,5(F9.5,2X))
253   280 FORMAT(1H ,7(F9.5,2X))
254       WRITE(6,290)
255   290 FORMAT(1H0,5X,'(NUMBER OF ABSORBED ENERGY PARTICLES)'/)
256       WRITE(6,250) NRD(2,7),NRD(3,7),NRD(4,7),NRD(5,7),NRD(6,7)
257       DO 1030 I=6,2,-1
258         WRITE(6,260) NRD(1,I),NRD(2,I),NRD(3,I),NRD(4,I),NRD(5,I),
259      *               NRD(6,I),NRD(7,I)
260  1030 CONTINUE
261       WRITE(6,250) NRD(2,1),NRD(3,1),NRD(4,1),NRD(5,1),NRD(6,1)
262       WRITE(6,300)
263   300 FORMAT(1H0,5X,'(RELATIVE ABSORBED ENERGY PROFILE)'/)
264       WRITE(6,270) RD(2,7),RD(3,7),RD(4,7),RD(5,7),RD(6,7)
265       DO 1040 I=6,2,-1
266         WRITE(6,280) RD(1,I),RD(2,I),RD(3,I),RD(4,I),RD(5,I),
267      *               RD(6,I),RD(7,I)
268  1040 CONTINUE
269       WRITE(6,270) RD(2,1),RD(3,1),RD(4,1),RD(5,1),RD(6,1)
270       STOP
271       END
272 *
273 **********************************************************************
274 *     RANDOM NUMBER GENERATOR
275 **********************************************************************
276       SUBROUTINE RANDOM(RAN,RAND)
277       REAL*8 RAND
278       RAND=DMOD(RAND*131075.0D0,2147483649.0D0)
279       RAN=SNGL(RAND/2147483649.0D0)
280       RETURN
281       END
```

FIG. 3.23. (*Continued*)

conditions as Fig. 3.15, the analytical results in Fig. 3.21 correspond to the absorption distribution in Fig. 3.15.

Figure 3.22 presents the analytical results for the case in which the values of (DXG, DYG) in line 12 of the RAT1 program listing in Fig. 3.16 are changed from (1.0, 1.0) to (5.0, 5.0) and the system size is enlarged by five times both longitudinally (in the x direction) and transversely (in the y direction). Because the size of each gas element is enlarged by five times accordingly, the fraction of self-absorption of the central gas element is approximately doubled, from 0.34202 to 0.75365, compared with Fig. 3.20 under the same gas absorption coefficient and wall emissivity. In contrast,

3.3. SIMULATION OF RADIATIVE HEAT TRANSFER

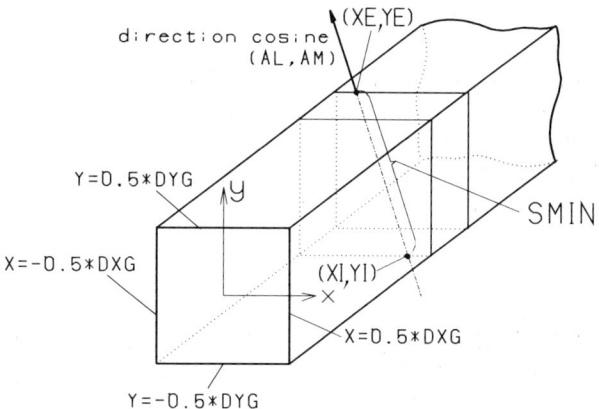

FIG. 3.24. Flight distance of energy particles SMIN through the rectangular gas element

the fraction of the radiative energy emitted outside of the gas element (4,4) and absorbed by the wall element (4,1), R_d (4,4,4,1), falls from 0.01047 in Fig. 3.20 to 0 in Fig. 3.22, resulting from the five times enlargement in the gas layer. This means that the radiative energy emitted from the central gas element fails to reach the wall surface.

3.3.5.2. Non-uniform Gas Absorption Coefficient Case

Figure 3.23 lists the program RAT2. It determines the absorption distribution of radiative energy in the same system of Fig. 3.9, but with a nonuniform gas absorption coefficient. Because the gas absorption coefficient differs in each gas element, the analysis of energy transmission in the gas phase requires the determination of the flight distance S_i of energy particles through each gas element, as stated in Section 3.3.1. In Fig. 3.23 S_i is expressed as the variable SMIN. The system of Fig. 3.9 is two dimensional. Because radiative energy propogates three dimensionally, each gas element in Fig. 3.9 extends infinitely in the direction perpendicular to the paper. It is appropriate to consider a rectangular slab of width DXG, height DYG, and infinite length, as shown in Fig. 3.24. For the computation of SMIN, a local coordinate system (X, Y) is established with its origin fixed at the center of each gas element. Let (XI, YI) be the coordinates for the incident point of radiative energy into the gas element, and (AL, AM) be the directional cosines in the x and y directions for expressing the traveling direction of an energy particle. Utilizing the

82 FORMULATION

information on (XI, YI) and (AL, AM), one can calculate the value of SMIN and the coordinates of the departure point (XE, YE) for the energy particle to exit from the gas element. Equation (3.17) determines the gas element that absorbs each energy particle. In other words, the products of SMIN [S_i in Eq. (3.17), flight distance in each gas element on the locus of each particle] and AK [K_i in Eq. (3.17), gas absorption coefficient of the corresponding gas element] of all gas elements along the locus of particle

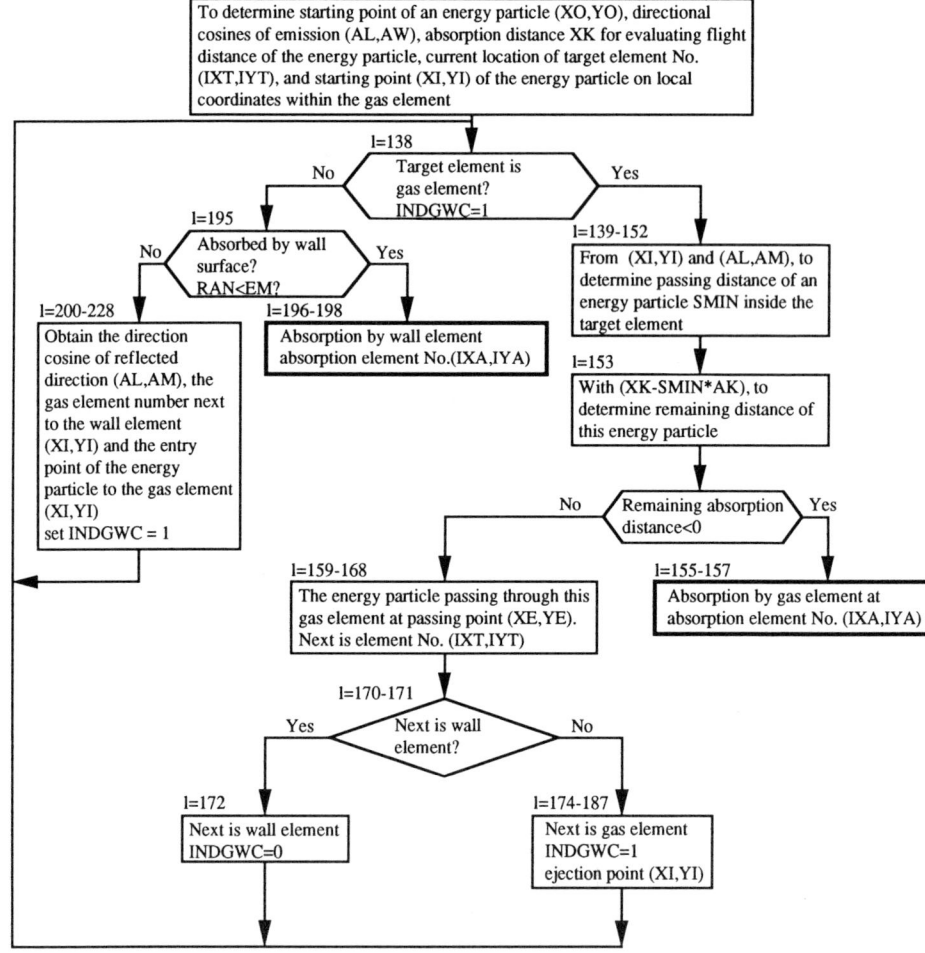

FIG. 3.25. Flowchart of principal part of program RAT2 (lines 64 through 228) of Fig. 3.23

3.3. SIMULATION OF RADIATIVE HEAT TRANSFER

flight are summed up. The energy particle is absorbed by the gas element at which the total sum of the products exceeds *KS* (the variable name in the program is XK), the absorption distance of the energy particle evaluated by Eq. (3.16).

Figure 3.25 shows the flowchart of the principal part of the RAT2 program shown in Fig. 3.23. The variables used in RAT2 are listed at the end of this monograph. In the RAT1 program, the gas absorption coefficient is uniform throughout the system and its value is directly read into the program by a READ statement. However, if it is necessary for the gas absorption coefficient to differ in each gas element, their values are specified by the DATA statement, as shown in lines 12 through 18 in Fig. 3.23. The gas space is divided into two domains with different gas absorption coefficients: 0.1 and 0.7 m^{-1}. Values of other variables are read into RAT2 via line 26, which states the wall emissivity EM, number of ejected energy particles NRAY, and identification of the ejecting element (IX, IY).

The printouts from the program are presented in Figs. 3.26 and 3.27. The first printout is the listing GAS ABSORPTION COEFF, the gas

```
***  MONTE CARLO SIMULATION OF RADIATIVE TRANSFER  ***

GAS ABSORPTION COEFF. (1/m)
  -1.00000   -1.00000   -1.00000   -1.00000   -1.00000   -1.00000   -1.00000
  -1.00000     .10000     .10000     .10000     .10000     .10000   -1.00000
  -1.00000     .10000     .10000     .10000     .10000     .10000   -1.00000
  -1.00000     .10000     .70000     .70000     .70000     .10000   -1.00000
  -1.00000     .10000     .70000     .70000     .70000     .10000   -1.00000
  -1.00000     .10000     .10000     .70000     .10000     .10000   -1.00000
  -1.00000   -1.00000   -1.00000   -1.00000   -1.00000   -1.00000   -1.00000
WALL EMISSIVITY              .10000E+01
NUMBER OF ENERGY PARTICLES    100000
EMITTING ELEMENT              (4,4)
ELEMENT SIZE        DXG=    1.00000(m)
                    DYG=    1.00000(m)


      (NUMBER OF ABSORBED ENERGY PARTICLES)

            1613      2545      3098      2648      1670
   1517      306       484       578       489       296      1611
   1686      349       806      1506       765       343      1699
   1552      232      7342     34061      7253       218      1509
   1235      173      3191      7393      3303       204      1119
    611      115       191      1227       178       105       654
             651      1062       749      1025       638


      (RELATIVE ABSORBED ENERGY PROFILE)

           .02446    .03860    .04698    .04016    .02533
  .02301   .00464    .00734    .00877    .00742    .00449    .02443
  .02557   .00529    .01222    .02284    .01160    .00520    .02577
  .02354   .00352    .11135    .34061    .11000    .00331    .02288
  .01873   .00262    .04839    .11212    .05009    .00309    .01697
  .00927   .00174    .00290    .01861    .00270    .00159    .00992
           .00987    .01611    .01136    .01554    .00968
```

FIG. 3.26. Output of RAT2 program, example 1

84   FORMULATION

```
*** MONTE CARLO SIMULATION OF RADIATIVE TRANSFER ***
GAS ABSORPTION COEFF. (1/m)
   -1.00000  -1.00000  -1.00000  -1.00000  -1.00000  -1.00000  -1.00000
   -1.00000    .10000    .10000    .10000    .10000    .10000  -1.00000
   -1.00000    .10000    .10000    .10000    .10000    .10000  -1.00000
   -1.00000    .10000    .70000    .70000    .70000    .10000  -1.00000
   -1.00000    .10000    .70000    .70000    .70000    .10000  -1.00000
   -1.00000    .10000    .10000    .70000    .10000    .10000  -1.00000
   -1.00000  -1.00000  -1.00000  -1.00000  -1.00000  -1.00000  -1.00000
WALL EMISSIVITY              .10000E+01
NUMBER OF ENERGY PARTICLES   100000
EMITTING ELEMENT             (1,4)
ELEMENT SIZE                 DXG=  1.00000(m)
                             DYG=  1.00000(m)


           (NUMBER OF ABSORBED ENERGY PARTICLES)
            2649    5220    4722    3326    2109
       0     571    1147     949     698     516    3323
       0    2544    2135    1091     610     363    2133
       0    9344   16669    4191    1346     108     630
       0    2633   10610    3283    1118      83     562
       0     583     914    2255     116      87     587
            2705    4741    2103     718     508

           (RELATIVE ABSORBED ENERGY PROFILE)
            .02649  .05220  .04722  .03326  .02109
 .00000     .00571  .01147  .00949  .00698  .00516   .03323
 .00000     .02544  .02135  .01091  .00610  .00363   .02133
 .00000     .09344  .16669  .04191  .01346  .00108   .00630
 .00000     .02633  .10610  .03283  .01118  .00083   .00562
 .00000     .00583  .00914  .02255  .00116  .00087   .00587
            .02705  .04741  .02103  .00718  .00508
```

FIG. 3.27. Output of RAT2 program, example 2

absorption coefficient of each gas element arranged in the same order as was read into the program under the data statement. The locations corresponding to the wall elements are given a value of -1.00000. This figure is an example of the present analysis. There are seven gas elements having an absorption coefficient of 0.7 m^{-1} that are located at and below the center of the system, and the remaining gas elements have an absorption coefficient of 0.1 m^{-1}. The two sets of outputs that follow, as in the RAT1 case, present the distribution of the number of absorbed energy particles, and that of the absorption fraction/READ value. In the case of Fig. 3.20, the absorption distribution of radiative energy emitted from the central gas element is uniform in the circumferential direction around the emitting element. In the RAT2 case, Fig. 3.26 shows that, with a distribution in the gas absorption coefficient, the radiative energy released from the central gas element (4,4) is largely absorbed by the gas elements (3,4), (4,5), and (5,4) of high gas absorption coefficient located to the left, right, and below it. Only a minor portion of the energy is absorbed by the gas element (4,3) of low gas absorption coefficient, which is located immediately over the central element. The situation is reflected in the analytical

3.3. SIMULATION OF RADIATIVE HEAT TRANSFER

results on the READ value that the gas elements (3,4), (4,5), and (5,4) have R_d (4,4,3,4) = 0.11135, R_d (4,4,4,5) = 0.11212, and R_d (4,4,5,4) = 0.11000 in contrast to the gas element (4,3) having R_d (4,4,4,3) = 0.02284, a factor of about 5 to 1. Now, compare the READ values of the wall elements (4,1) and (4,7), which are located above and below the central gas element, respectively. It is R_d (4,4,4,1) = 0.04698 to R_d (4,4,4,7) = 0.01136; that is, the upper wall element absorbs four times more than the lower one. This is because the gas element below the central one has a higher gas absorption coefficient than the one above. Accordingly, the intensity of the radiative energy released from the central gas element is substantially diminished by absorption in the course of its transmission before reaching the lower wall element (4,7).

Under an identical distribution of the gas absorptivity, Fig. 3.27 presents the analytical results for the absorption distribution of radiative energy released from the left-side wall element (1,4). With regard to the amount of energy absorbed by each element on the right-side surface, the wall elements in the upper half absorb about 5.5 times more than those in the lower half, because the energy particles travel through the domain of low gas absorption coefficients.

Chapter 4
Methods of Solution

In a combined radiative and convective system, as depicted in Fig. 3.1 of Section 3.2, two methods are available to determine the gas temperature distribution and the wall heat flux. They are the energy method and the radiative energy absorption distribution (READ) method.

4.1. Energy Method

Figure 4.1 is a flowchart used for computations in the energy method, in which e_0 denotes the radiative energy that each energy particle transmits. Part (a) of the flowchart is the part that uses the procedure described in Section 3.3.5 to determine the number of energy particles N_a being absorbed by each element. The rate of radiative energy being absorbed by each element, $q_{r,\text{in}}$, can be obtained by multiplying N_a by e_0:

$$q_{r,\text{in}} = N_a \cdot e_0. \tag{4.1}$$

This method requires a lengthy computational time because the time-consuming part [part (a) in the flowchart] of the Monte Carlo method is in the temperature convergence loop, which requires repeated computations until temperature convergence is achieved. In this method, the Monte Carlo method is utilized to evaluate directly the rates of radiative energy that is incident on and is absorbed by each element. In Eq. (3.1) $q_{r,\text{out}}$ varies each time the temperature distribution changes with a repeated computation, requiring repetition of the Monte Carlo computation.

4.2. READ Method

To reduce the computational time of the energy method, the READ method is proposed to avoid repeating the computational part in the

4.2. READ METHOD

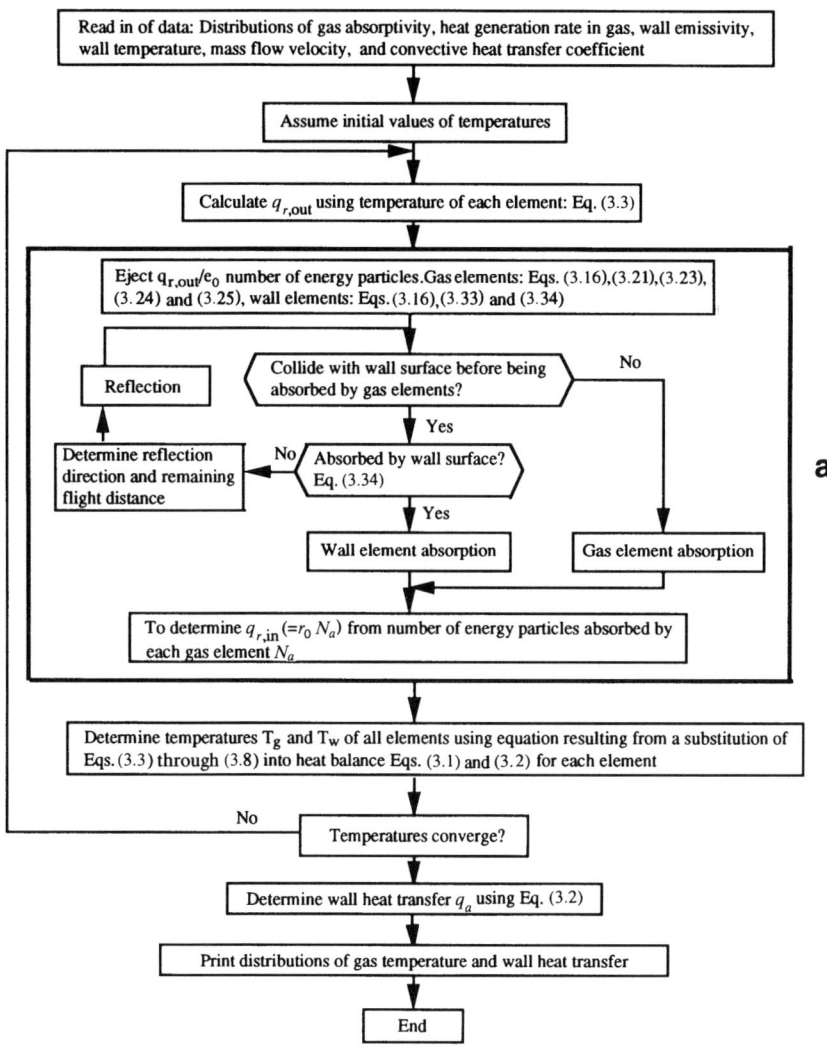

FIG. 4.1. Flowchart for radiative heat transfer analysis, (a) by means of the energy method, which uses the Monte Carlo technique

Monte Carlo method. The new scheme realizes that the magnitude of R_d in Eq. (3.5) depends only on the system geometry and the distribution of radiative physical properties, K (absorption coefficient of each gas element) and ε (emissivity of wall elements), as mentioned in Section 3.2. It follows the procedure of Section 3.3.5 to calculate the magnitude of R_d

only once by means of the Monte Carlo method and then determine $q_{r,\text{in}}$ using Eq. (3.5).

The R_d is expressed as the function $R_d[(i_0, j_0, k_0), (i_i, j_i, k_i)]$ using the locations of the emitting element (i_0, j_0, k_0) and the absorbing element (i_i, j_i, k_i) as the indices. It represents the fraction of the radiative energy absorbed by each of the other elements to the net radiative energy being released from an emitting element, which is the difference between the total emitted and the self-absorption. Here, i, j, and k denote the identification numbers of each element in a three-dimensional system. The total summation of the R_d's, representing the fractions of the radiative energy released from the emitting element (i_0, j_0, k_0) being absorbed by all other elements, is equal to unity, according to this definition:

$$1 = \sum_l \sum_m \sum_n R_d[(i_0, j_0, k_0), (i_l, j_m, k_n)]. \qquad (4.2)$$

The magnitude of one set of R_d's is constant, as long as there is no change in the geometrical relationship between each element and the distribution of radiative physical properties in the system. This R_d is called the radiative energy absorption distribution, READ, and can be determined by the Monte Carlo method as follows. Furthermore, the magnitude of the self-absorption fractions, $\alpha_{s,g}$ and $\alpha_{s,w}$, in Eqs. (3.3) and (3.4) can be evaluated in the course of computing the READ.

Recall that, in Section 3.3.5, the programs RAT1 and RAT2 are used to determine the distribution of absorptive locations of NRAY number of energy particles ejected from each element. To utilize the results to evaluate the self-absorption fractions, $\alpha_{s,g}$ and $\alpha_{s,w}$, and the magnitude of READ, R_d, the following computations are performed in the RAT1 and RAT2 programs. Let N_a be the sum of the number of energy particles absorbed within the emitting element before they get out of the element (in case the emitting element is a gas element) and the number of energy particles that have escaped from the emitting element, but are eventually absorbed by the element due to a change in the flight directions caused by reflection, diffuse, etc. Then, the self-absorption ratio of the emitting element is

$$\alpha_{s,i}(i_0, j_0, k_0) = N_a/\text{NRAY} \qquad \text{for } (i = g, w). \qquad (4.3)$$

If N_1 is the number of energy particles being absorbed by other elements (i_1, j_1, k_1), then the magnitude of the READ between the emitting and absorbing elements is

$$R_d[(i_0, j_0, k_0), (i_1, j_1, k_1)] = N_1/(\text{NRAY} - N_a). \qquad (4.4)$$

Because this READ is defined between the emitting element and all other elements, the number of READs is $n(n-1)$ where n is the number of total elements. In the energy method, the number of energy particles ejected from each element is proportional to $q_{r,\text{out}}$, the radiative energy emitted from the element. It is equal to NRAY for all elements in the READ method. The reason is that the method determines the rate of radiative heat transmission from the elements a to b, as seen in Eq. (3.5), as

$$R_d(a, b) \cdot q_{r,\text{out}}(a).$$

It is the product of two irrelevant quantities, the radiative energy from the irradiating element a, $q_{r,\text{out}}(a)$, and the fraction of radiative heat exchange between the elements, $R_d(a, b)$. The magnitude of R_d can be obtained, independently, unrelated to the rate of radiative heat from the emitting element $q_{r,\text{out}}$. Because the use of an excessively small number of energy particles would result in an increase in data scattering, it is desirable to test a program by varying the particle number, NRAY, to make sure that the resulting data scatterings are within a specific range.

Figure 4.2 is a flowchart for radiative heat transfer analysis by means of the READ method. Because the part for computing the READ value by the Monte Carlo method, part (a), is outside the computational loop for temperature convergence, the time-consuming computation by the Monte Carlo method is performed only once, resulting in a reduction in computational time compared with the energy method. The section names indicated on the left side of the flowchart correspond to the main routine of the program RADIAN to be introduced in Section 6.2.

Thus far, the energy and READ methods have been introduced as the means by which to apply the Monte Carlo technique in radiative heat transfer analyses. Next we compare the computational time and memory usage in both methods. Let us first examine the computational time. Both methods consume most of their computational time in calculating the absorption distribution of radiative energy by means of the Monte Carlo technique, that is, part (a) of Figs. 4.1 and 4.2. Hence, it is obvious that the entire computational time would be shortened by reducing the computational time of part (a). Because the energy method has part (a) within the loop of calculating temperature convergence, as illustrated in Fig. 4.1, whereas the READ method has it outside of the loop, as seen in Fig. 4.2, the former would have to spend more time than the latter in repeating the computations for temperature convergence. In practice, the computations for temperature convergence are repeated 5 to 10 times, implying that the computation time of the energy method is 5 to 10 times that of the READ method.

As far as memory usage is concerned, the energy method deals with the number of energy particles being absorbed by each element, therefore, the

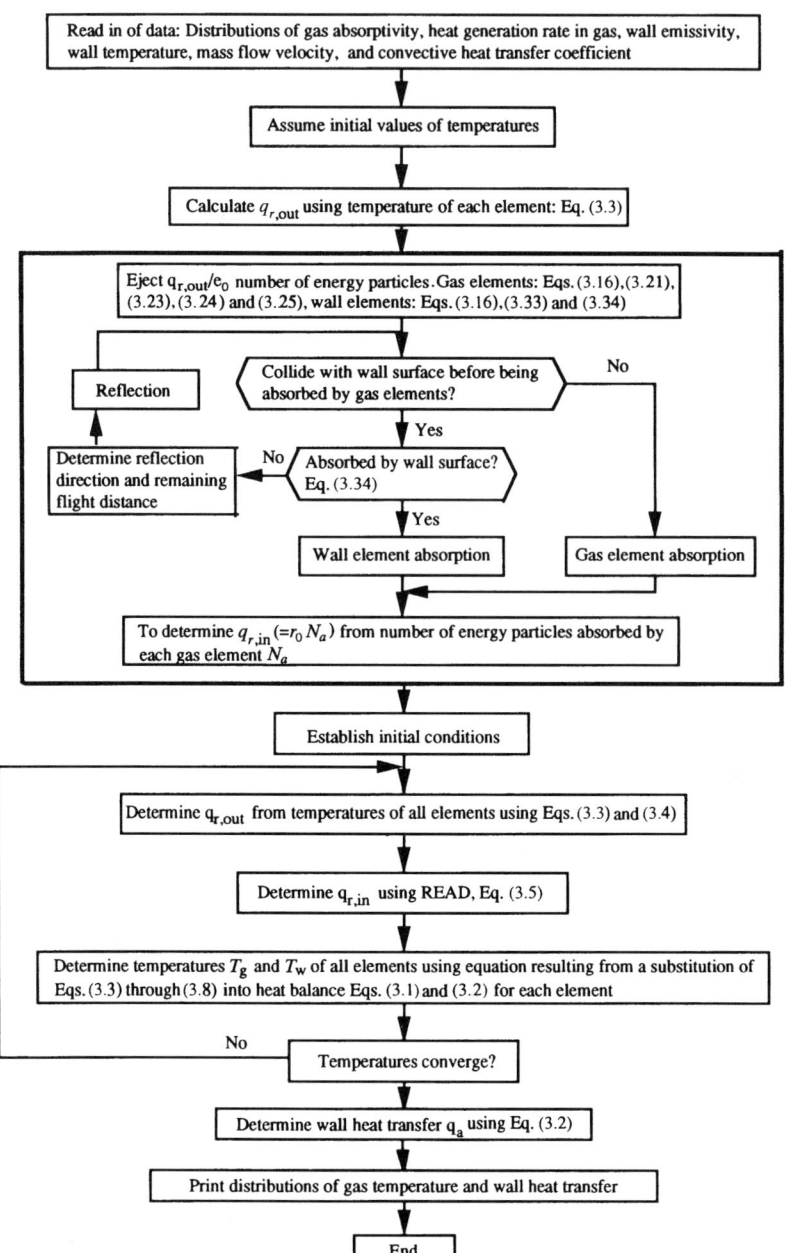

FIG. 4.2. Flowchart for radiative heat transfer analysis, (a) by means of the READ method, which uses the Monte Carlo technique

number of its memory usage is the element number n. In contrast, the number of its memory usage in the READ method includes the number of gas elements pertinent to their self-absorption fractions, Equation (4.3), and the number to memorize the READ values in Eq. (4.4), which has the number of both the emitting and absorbing elements as the index. Its necessary number is, therefore, $n + n(n - 1) = n^2$. One should bear in mind that the READ method may run out of memory with an increase in the element subdivisions in a multiple-dimensional case.

In the practical application of radiative heat transfer, a fine element subdivision is not needed except for studies of the local structure of flames, boundary layer phenomena, etc. This is because radiative heat transfer is an integral phenomenon and the variations in a small domain usually have little effect on the result. On the contrary, a relatively coarse element subdivision is quite often sufficient and the cases are generally abundant in which radiative analysis is conducted by means of the READ method. However, it is necessary to employ the energy method in the cases of a large number of element subdivisions and a shortage of computer memory capacity.

Chapter 5
Special Treatises

5.1. Introduction

One may consider radiation heat transfer between two or more bodies in three steps. The first step is the simplest case, where the space between these radiating surfaces is either a vacuum or is filled with gases or gas mixtures having symmetrical molecules. These gases or gas mixtures are practically transparent to thermal radiation; they neither emit nor absorb appreciable amounts of radiant energy. Dry air, O_2, N_2, H_2, etc., belong in this category. This case will be treated in Section 6.3. In the second step, the space between the radiating bodies is filled with heteropolar gases and vapors such as CO_2, H_2O, SO_2, CO, NH_3, hydrocarbons, and alcohols. These gases and vapors emit and absorb radiation between narrow regions of wavelength called bands. They are of importance in heat transfer equipment. In particular, H_2O and CO_2 are the most important of the gases in furnaces. This case will be presented in Section 6.2, in which the calculation of the radiation emitted or absorbed by a gas layer involves its temperature, pressure, shape, and volume.

The third step is to treat the case in which gases or vapors between the radiating bodies are seeded with particles, including solid particles and liquid droplets. Scattering is any encounter between a photon and one or more particles. During an encounter, the photon does not lose all of its energy, but may undergo a change in direction and a partial loss or gain of energy.

The scattering can be classified into two categories: elastic and inelastic. The energy of the photon is unchanged in elastic scattering, but changed in inelastic scattering. In engineering applications, most scattering events of importance are elastic or very nearly so.

In addition to the special treatise on scattering by particles discussed here, another one dealing with nonorthogonal boundary cases is presented in this chapter. The system under consideration may take an irregular geometry which needs special treatment. Section 5.3 addresses such problems.

5.2. Scattering by Particles

The preceding chapters treat a combined radiation convection heat transfer in a system with only absorption (no scattering in the gas phase). This section deals with the analytical method for the case in which solid particles or liquid droplets (both to be referred to as particles) act as scattering sources in the gas phase and a temperature difference exists between the particles and the gas. Section 1.4 defines the physical quantities that are needed in expressing the optical characteristics of the particle-containing gas. Figure 5.1 depicts the flowchart used to analyze such a system using the Monte Carlo technique. The analytical method is presented next.

The radiative heat transfer inside a particle-containing gas surrounded by solid walls is the absorption distribution of radiative energy emitted from the gas, particles, and solid walls, in proportion to the fourth power of each temperature. The physical quantities being prescribed in the problem include the distribution of particle concentrations in the gas, mass velocity distributions in the gas and particles, optical properties of the particle-containing gas as defined in Section 1.4, heat generation distribution in the gas or particles resulting from combustion of gaseous fuel or solid fuel, boundary conditions of the surrounding solid walls (either temperature or heat flux), and emissivity of the solid walls. The objectives of the analysis are the temperature distributions of the gas and particles and the heat flux or temperature distribution of the solid walls. Hence, it is necessary to treat the gas and particles separately. For the sake of defining the distribution of identification variables for the particles, the particle elements are needed in addition to the gas and wall elements. An example is presented in Fig. 5.2. For simplicity, it is better to subdivide the particle elements that are identical to the gas elements. With the particle and gas elements in the same shape, the ejecting direction and location of energy particles from each corresponding element can be determined by Eqs. (3.23) and (3.24) and Eqs. (3.25)–(3.27), respectively. Furthermore, the ejection of energy particles from the wall elements is, as in Section 3.3.3, from an arbitrary point of the element in the direction described by Eqs. (3.34) and (3.35). During the flight of energy particles ejected from each element through the particle-containing gas, a distance of S and the attenuation coefficient of the particle-containing gas a can be determined by the equation

$$aS = -\ln(1 - R_s), \qquad (5.1)$$

where aS is called the *attenuation distance* of each energy particle. This equation can also be obtained from Eq. (1.33) in a manner similar to how KS in Eq. (3.16) for an absorbing gas was derived from Eq. (1.25).

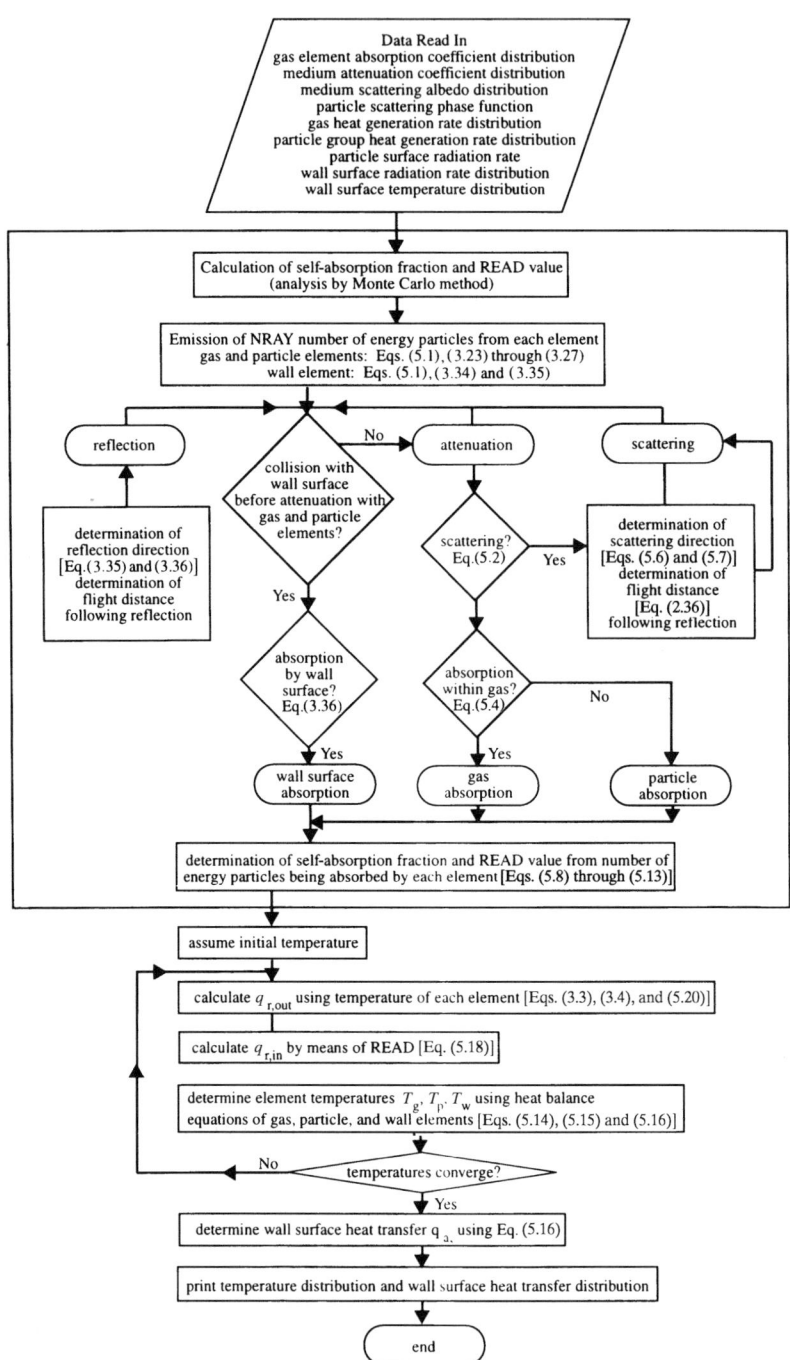

FIG. 5.1. Flow diagram for radiative heat transfer analysis in absorbing and scattering media

5.2. SCATTERING BY PARTICLES

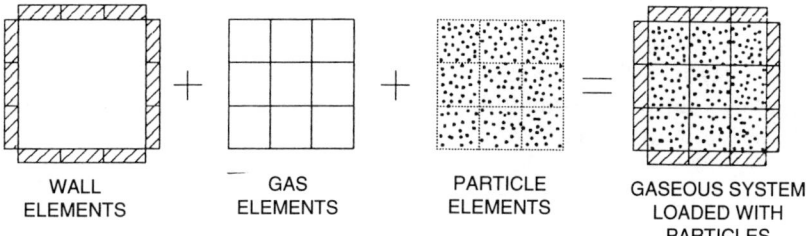

FIG. 5.2. Elements in a gas system loaded with particles

After flying the attenuation distance of aS, the energy particle is either absorbed by the gas and its containing particles or scattered by the particles. Whether each energy particle is absorbed or scattered is determined by the diffuse albedo defined by Eq. (1.37) ω and the uniform random number for scattering albedo R_ω: The energy particle is scattered when

$$R_\omega \leq \omega \tag{5.2}$$

and is absorbed by the gas or its containing particles if

$$R_\omega > \omega. \tag{5.3}$$

In the absorption case, *whether the absorption of an energy particle is by the gas or by its containing particles* is decided by the gas absorption coefficient K, absorption cross section of particle groups, and uniform random number for absorption R_a.
The absorption is by the gas when

$$R_a \leq K/(K + \sigma_a), \tag{5.4}$$

whereas the absorption is by the particle groups if

$$R_a > K/(K + \sigma_a). \tag{5.5}$$

The energy particle that meets the condition of Eq. (5.2) is scattered in the direction (θ, ϕ), in Fig. 1.8, which follows the distribution of the scattering phase function ϕ defined in Section 1.4.3. The phenomenon is simulated by means of the Monte Carlo method. The scattered direction of each energy particle is obtained by the inverse transformation method using the cumulative scattering probability, i.e., the probability for the scattering angle to be within the range of $0-\theta$ or $0-\phi$. When this method is applied with respect to the scattering phase function defined by Eq. (1.39), one obtains

$$\frac{\theta}{2} - \frac{3}{8}\sin 2\theta + \frac{\theta \cos 2\theta}{4} - \frac{3\pi}{4}R_{\theta s} = 0. \tag{5.6}$$

With a substitution of a value between 0 and 1 into the uniform random number $R_{\theta s}$, Eq. (5.6) is solved for θ, which is the scattered direction of each energy particle. The scattered characteristics of the energy particle that follows Eq. (1.39) is uniform in the ϕ direction. Hence, the scattering angle in the ϕ direction is found to be

$$\phi = 2\pi R_{\phi s}. \tag{5.7}$$

The flight distance following the scattering can be obtained from Eq. (5.1), similar to the emission of energy particles. One can thus obtain the distribution of absorption points of energy particles that are ejected from each gas, particle, and wall element. As mentioned previously, the gas and particle elements are identical in shape, and the equations for evaluating the ejection point, direction, and attenuation distance of energy particles from both elements are also identical. Hence, it is unnecessary to conduct separate analyses on the distribution of absorption points of energy particles for both elements. The gas and particle elements are handled as one in the flowchart, Fig. 5.1.

In a manner similar to Section 4.2, the self-absorption fraction of each element and the READ value can be obtained as follows. Out of NRAY number of energy particles ejected from the gas particle elements, let N_{ag} be the number being absorbed by the emitting gas elements and N_{ap} be that by the emitting particle elements. Like Eq. (4.3), the self-absorption ratios of the emitting gas and particle elements are, respectively,

$$\alpha_{ag} = N_{ag}/\text{NRAY}, \tag{5.8}$$

$$\alpha_{ap} = N_{ap}/\text{NRAY}. \tag{5.9}$$

Similarly, the self-absorption ratio of the wall elements that eject NRAY number of energy particles is

$$\alpha_{sw} = N_{aw}/\text{NRAY}. \tag{5.10}$$

Here, N_{aw} is the number of energy particles being absorbed by the emitting wall elements. Let N_1 be the number of energy particles being absorbed by the absorbing elements. As mentioned in Section 4.2, the READ values are determined as, similar to Eq. (4.4),

$$R_d(\text{emitting gas element, absorbing element}) = N_1/(\text{NRAY} - N_{ag}), \tag{5.11}$$

$$R_d(\text{emitting particle element, absorbing element}) = N_1/(\text{NRAY} - N_{ap}), \tag{5.12}$$

5.2. SCATTERING BY PARTICLES

$$R_d(\text{emitting wall element, absorbing element}) = N_1/(\text{NRAY} - N_{aw}). \tag{5.13}$$

These computations take place in the upper half of the flowchart in Fig. 5.1.

Next we explain the lower half of the flowchart. This portion calculates the temperature distribution in each element using the self-absorption fraction and the READ value obtained by the upper half. However, this analysis requires the following heat balance equations for the three elements:

Heat balance for gas elements:

$$q_{r,\text{out},g} + q_{c,gw} + q_{c,gp} + q_{f,\text{out},g} = q_{r,\text{in},g} + q_{h,g} + q_{f,\text{in},g}, \tag{5.14}$$

heat balance for particle elements:

$$q_{r,\text{out},p} + q_{c,pw} + q_{f,\text{out},p} = q_{r,\text{in},p} + q_{c,gp} + q_{h,p} + q_{f,\text{in},p}, \tag{5.15}$$

and heat balance for wall elements:

$$q_{r,\text{out},w} + q_a = q_{r,\text{in},w} + q_{c,gw} + q_{c,pw}. \tag{5.16}$$

The LHS terms in the three equations denote the out-flow components of each corresponding element, where the RHS terms are the in-flow heat components. Equation (5.14) corresponds to Eq. (3.1). The $q_{r,\text{out},g}$, $q_{c,gw}$, $q_{c,gp}$, $q_{f,\text{out},g}$, $q_{r,\text{in},g}$, $q_{f,\text{in},g}$ terms represent, respectively, the out-flow radiative heat, convective heat with the wall surface, convective heat with the particles, out-going enthalpy flow, in-flow radiative heat, and in-coming enthalpy flow of the gas elements. They can be determined by Eqs. (3.3), (3.6), (5.17), (3.7), (5.18), and (3.8), respectively. The $q_{h,g}$ term in Eq. (5.14) denotes the rate of heat generation within the gas element. The rate of heat transfer from gas to particles within an element can be expressed as

$$q_{c,gp} = h_{gp} \pi d^2 N \Delta V (T_g - T_p). \tag{5.17}$$

Here, h_{gp} represents the gas–particle convective heat transfer coefficient. The rate of radiative heat absorption by each element is determined by

$$q_{r,\text{in},i} = \underbrace{\sum R_d \cdot q_{r,\text{out},g}}_{\text{gas}} + \underbrace{\sum R_d \cdot q_{r,\text{out},p}}_{\text{particle}} + \underbrace{\sum R_d \cdot q_{r,\text{out},w}}_{\text{wall}}, \tag{5.18}$$

in which $i = g, p, w$. This equation corresponds to Eq. (3.5). Each term on the RHS is a summation of radiative heats that are transferred to the element i from each of the gas, particle, or wall elements.

Equation (5.15), which takes the same form as Eq. (5.14), is the heat balance equation for the particle elements. Each term in the expression can be expressed as follows:

$$q_{r,\text{out},p} = (1 - \alpha_{s,p})\varepsilon_p \sigma T_p^4 \pi d^2 N \Delta V. \tag{5.19}$$

With a substitution of Eq. (1.35), it can be rewritten as

$$q_{r,\text{out},p} = 4(1 - \alpha_{s,p})\sigma T_p^4 \sigma_a \Delta V, \tag{5.20}$$

where $q_{c,pw}$ is the convective heat transfer rate induced by the impact of particles against the solid walls. It is

$$q_{c,pw} = h_{pw} \Delta S (T_p - T_w), \tag{5.21}$$

wherein h_{pw} denotes the particle–wall heat transfer coefficient and ΔS is the corresponding convective heat transfer area. The $q_{f,\text{out},p}$ and $q_{f,\text{in},p}$ terms are, respectively, the outgoing and incoming components of enthalpy flow for the particle element, which is induced by the migration of particles. They can be determined by

$$q_{f,\text{out},p} = W_p C p_p T_p \Delta S_q, \tag{5.22}$$

$$q_{f,\text{in},p} = W_p C p_p T_{p,\text{up}} \Delta S_q, \tag{5.23}$$

which corresponds to Eqs. (3.8) and (3.7), respectively. Here, W_p signifies the mass flow rate of particles; Cp_p, specific heat; and T_p, temperature. The subscript "up" refers to the upstream element. The $q_{r,\text{in},p}$ and $q_{c,gp}$ terms on the RHS on Eq. (5.15) can be determined by Eqs. (5.18) and (5.17), respectively, and $q_{h,p}$ represents the total heat generation rate of the particles within the particle element. Equation (5.16) corresponds to Eq. (3.2) in an absorptive gas system; $q_{r,\text{out},w}$, $q_{r,\text{in},w}$, and $q_{c,gw}$ terms can be determined by Eqs. (3.4), (5.18), and (3.6), respectively; and q_a denotes the net heat transfer of the wall element. It is either specified as a boundary condition or obtained from analysis. The $q_{c,pw}$ term is identical to the second term on the LHS of Eq. (5.15).

The preceding expressions are the basic equations used to determine the temperatures and wall heat fluxes. They are employed to express all heat component terms in Eqs. (5.14), (5.15), and (5.16) in terms of the element temperatures T_g, T_p, and T_w. Then, with the specification of the self-absorption fractions, READ value, all heat transfer coefficients, distribution of heat generation rate, mass velocity distributions of the gas and particles, and T_w or q_a as a boundary condition of the wall element, one can calculate the gas temperature T_g, particle temperature T_p, and wall heat flux q_a (with T_w specified as a wall boundary condition) or wall temperature T_w (with q_a specified as a wall boundary condition). These computa-

5.3. Nonorthogonal Boundary Cases

tions are performed by the temperature convergence loop in the lower half of the flow chart in Fig. 5.1.

5.3. Nonorthogonal Boundary Cases

In the previous sections, all systems consist of either rectangular (in two-dimensional cases) or brick-shaped (in three-dimensional cases) gas elements and the surrounding wall elements. However, there are cases in which the shape of the actual system is too complex to be expressed by these elements. This section describes the use of the cylindrical and Cartesian coordinate systems for an arbitrary boundary shape.

5.3.1. CYLINDRICAL COORDINATE SYSTEM

Figure 5.3 illustrates an example of element subdivision in the cylindrical coordinate system. The radiative heat transfer analysis of such a system by means of the Monte Carlo method is, in principle, similar to that of the preceding sections. That is, the same number of energy particles is ejected from each of the gas and wall elements, the elements that absorb these

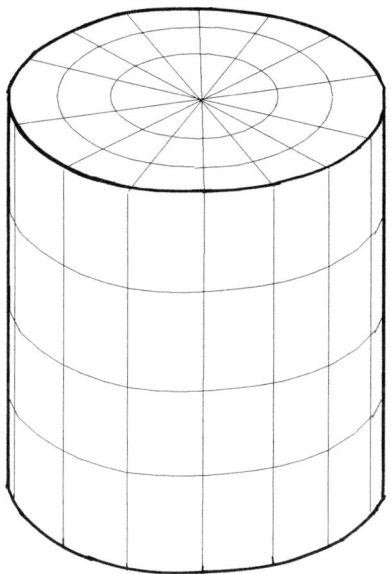

FIG. 5.3. Element subdivision in cylindrical coordinate system

energy particles are identified, and the distributions of the self-absorption fractions and READ values in the system are determined from the results. Referring to the explanation of Eqs. (3.25)–(3.27), the location of energy particles emitted from a gas element, $P(r, \theta, z)$, must be uniformly distributed in the gas element of the shape shown in Fig. 5.4. Let $f(r)\,dr$ be the probability for the r coordinate of the ejection point to be in the interval between r and $(r + dr)$. It is equal to the shaded area divided by the total area of the element (area enclosed by $abcd$) in Fig. 5.5. That is,

$$f(r)\,dr = \frac{2\pi r\,dr \cdot (\theta_2 - \theta_1)/2\pi}{\pi(r_2^2 - r_1^2) \cdot (\theta_2 - \theta_1)/2\pi}$$

$$= \left[2r/(r_2^2 - r_1^2)\right] dr. \quad (5.24)$$

An application of the inverse transformation method, described in Section 3.3.1, and taking into account $r_1 \leq r \leq r_2$, yields

$$\xi = \int_{r_1}^{r} f(\eta)\,d\eta = \frac{r^2 - r_1^2}{r_2^2 - r_1^2}. \quad (5.25)$$

By equating ξ to the uniform random number R_r, one obtains the r coordinate of the energy ejection point as

$$r = \left[(r_2^2 - r_1^2)R_r + r_1^2\right]^{1/2}. \quad (5.26)$$

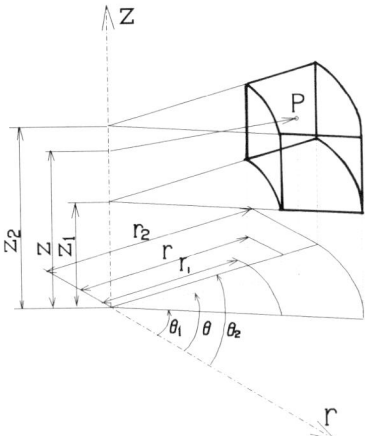

FIG. 5.4. Cylindrical coordinate system

5.3. NONORTHOGONAL BOUNDARY CASES

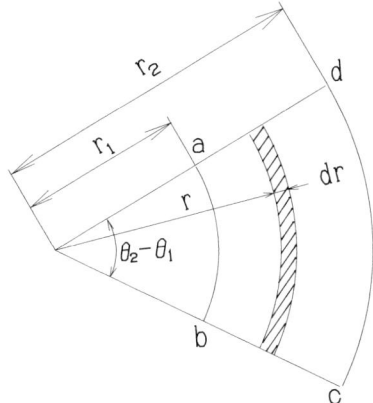

FIG. 5.5. Cylindrical coordinate system

Concerning the θ and z directions, energy particles are ejected proportionately in the θ_1 to θ_2 and z_1 to z_2 intervals. Hence, θ and z can be obtained, using the uniform random numbers R_θ and θ_z, as

$$\theta = (\theta_2 - \theta_1)R_\theta + \theta_1, \quad (5.27)$$

$$z = (z_2 - z_1)R_z + z_1. \quad (5.28)$$

Equations (5.26), (5.27), and (5.28) correspond to Eqs. (3.25)–(3.27) for the coordinates of the ejection point in the Cartesian coordinate system. Figure 5.6 shows the distribution of the ejection points of energy particles in a two-dimensional fan-shaped gas element, which are obtained through consecutive substitutions of a pair of the uniform random numbers (R_r, R_θ) into Eqs. (5.26) and (5.27). It is disclosed that by using Eqs. (5.26) and (5.27), the energy particles are uniformly ejected from the domain of a fan-shaped gas element.

The flight direction and distance of each energy particle are obtained by using Eqs. (3.23), (3.24), and (3.16), respectively.

Next we concern ourselves with wall elements. As in the preceding sections, each energy particle is ejected from an arbitrary position on each wall element in the direction of Eqs. (3.34) and (3.35) with its flight distance determined by Eq. (3.16). Furthermore, Eqs. (3.23), (3.24), (3.34), and (3.35) apply, with an appropriate definition of the local coordinate system of each element.

The procedure described in Section 3.3.5 can also be utilized to determine the absorbing elements of energy particles ejected from the gas and wall elements. An example of an analysis on the cylindrical coordinates will be presented in Part III, Section 7.3.

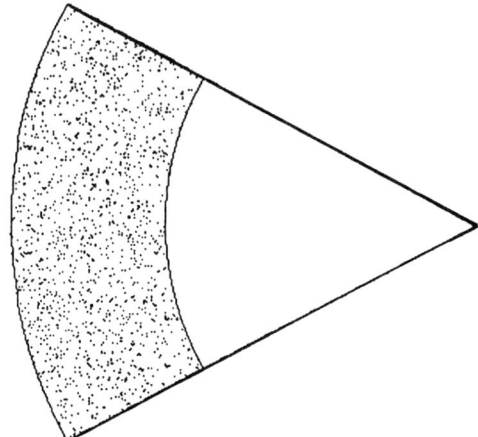

Fig. 5.6. Distribution of emitting points in cylindrical coordinates

5.3.2. Nonorthogonal Boundary

Consider radiative heat transfer within a system surrounded by a wall of arbitrary shape, as illustrated in Fig. 5.7. The rectangular gas elements whose sides are perpendicular to the coordinate axes are superimposed on the boundaries of the arbitrary shape. The gas and wall elements are given their respective identification numbers ISG(NG) and ISW(NW). Here, NG and NW are the integers to identify each gas and wall element, respectively. The gas elements with ISG = 1 in Fig. 5.7 are those rectangular gas elements that are either not in contact with any wall elements or in contact with the wall element of the same side length. Those gas elements in Fig. 5.7 that do not meet the conditions (ISG ≠ 1) are shown in Fig. 5.8. They are given the identification numbers ISG = 2, 3, 4, etc. The wall elements in contact with the gas elements of ISG = 1 are given ISW = 1. As shown

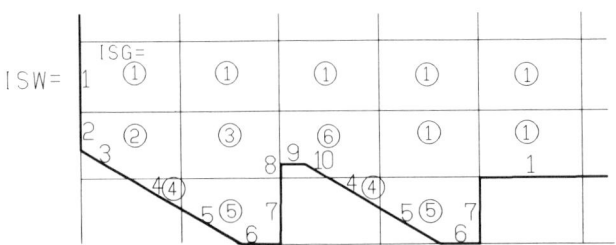

Fig. 5.7. An example of an nonorthogonal boundary

5.3. NONORTHOGONAL BOUNDARY CASES

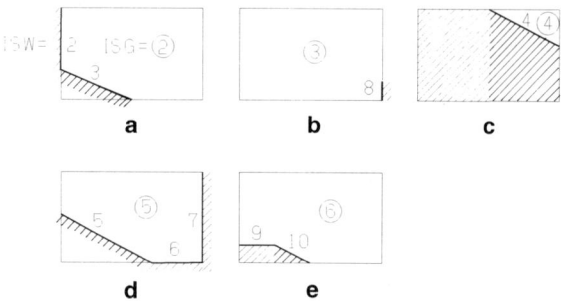

FIG. 5.8. Special boundary elements

in Fig. 5.7, the other wall elements are numbered ISW = 2, 3, 4, etc., with the same identification number for those of the same shape. The treatment of the (ISG = 1, ISW = 1) follows that of the rectangular element. The programs for the ejection, absorption, reflection, etc., of radiative energy particles must be written separately for each of the other elements (ISG ≠ 1, ISW ≠ 1). For example, the element (ISG = 2, ISW = 2, 3), i.e., case (a), in Fig. 5.8 is treated as follows: The ejection point of energy particles from the gas element must be uniformly distributed in the unhatched portion of the rectangle. It is equivalent to distributing uniformly point (x, y) inside a rectangle of the height Δy and width Δx in the local coordinate system, x versus y, and discarding those points distributed in the shaded triangle as depicted in Fig. 5.9. Each (x, y) point is determined using two uniform random numbers, R_x and R_y, as

$$x = \Delta x \cdot R_x, \tag{5.29}$$

$$y = \Delta y \cdot R_y. \tag{5.30}$$

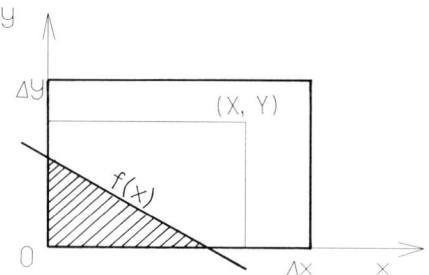

FIG. 5.9. A method to obtain uniformly distributed emitting points

Let the wall boundary traversing the gas element be $y = f(x)$. By discarding those (x, y) points that fail to meet the condition

$$y > f(x),$$

then all other (x, y) points obtained by Eqs. (5.29) and (5.30) would fall uniformly in the unhatched portion of the rectangle $(\Delta x, \Delta y)$ in Fig. 5.9.

Next the ejection direction and flight distance of energy particles are evaluated by Eqs. (3.23), (3.24), and (3.16), respectively. As in the preceding sections, a constant number of energy particles are ejected from arbitrary points on each wall element ISW = 2, 3 in the direction of Eqs. (3.34) and (3.35) and with the flight distance of Eq. (3.16).

Part III

APPLICATIONS OF THE MONTE CARLO METHOD

Chapter 6
Two-Dimensional Systems

6.1. Introduction

This chapter presents the applications of the READ method to analyze the combined radiative–convective heat transfer processes inside a radiant–absorptive gas enclosed by two-dimensional solid walls of arbitrary geometry. Two cases are treated, which are different in terms of the radiative characteristics of the medium (gas) enclosed between the heat-exchanging walls: absorbing-emitting gas and nonparticipating gas. Two different computer programs have been developed. One is called RADIAN for the absorbing-emitting gas case, and the other is RADIANW to be used in the nonparticipating gas case.

6.2. Radiative Heat Transfer in the Absorbing–Emitting Gas: Program RADIAN

A computer program, called RADIAN, was developed to analyze radiative heat transfer in a system with an absorbing–emitting gas. It subdivides a rectangular domain of any size and geometry into equal-sized, rectangular, gray gas elements, as shown in Fig. 6.1 The gas elements are bordered by equal-sized, gray wall elements, as indicated by the bold solid lines within which the combined radiant–convective heat transfer analysis is performed. In the figure, the hatch-lined elements are outside the system and belong to a domain that is excluded from the analysis. They are included for convenience in programming. Each element is numbered accordingly. Numbers 1 through 30 represent gas elements and numbers 1 through 26 wall elements in Fig. 6.1 The program listing for RADIAN is given in Fig. 6.2 RADIAN that can take up to 40 gas elements and 28 wall elements. For problems that require more elements, one simply changes the 40 and 28, which correspond to the indices of COMMON STATEMENTS and DIMENSION STATEMENTS, respectively, in lines 14

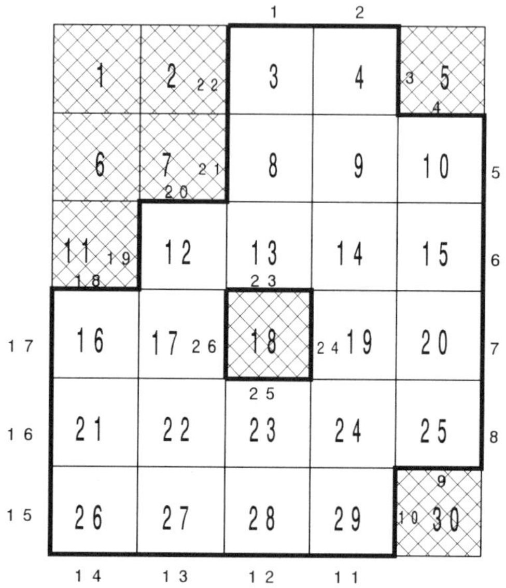

108   TWO-DIMENSIONAL SYSTEMS

FIG. 6.1. Example of mesh division of RADIAN program

through 21 in the program listing. However, in the case of 16-bit personal computers (PCs), such an increase in the element number may strain the PC's memory.

To define the domain of the gas elements involved in computations, we must enclose the adjacent wall elements, as shown by the numbers 1 through 22 and 23 through 26 of Fig. 6.1. Should there be an opening in the walls, a fictitious wall surface having an identical transmissive characteristic for radiative energy must be substituted for the opening. For instance, if the wall surfaces have an opening into the atmosphere at room temperature, the radiant energy emitting out through the opening will not return to the system. Because the radiant energy from the surroundings entering the system interior is a thermal radiation corresponding to the room temperature, the opening can be replaced by a wall surface at room temperature and with an emissivity of unity, from the radiative viewpoint. However, as to the convective heat transfer, the opening should be considered a fictitious wall with zero convective heat transfer coefficient. In other words, the fictitious surface affects only the radiation, not the enthalpy transport by convection. It is considered a surface that exerts no resistance to the inlet or outlet flow of fluids. Accordingly, the corresponding wall boundary conditions should "read" a fictitious porous wall at room temperature, with unit emissivity, and having zero resistance to flows.

6.2. RADIATIVE HEAT TRANSFER: PROGRAM RADIAN

Another example concerns an opening connected to a long duct, such as a smoke duct at the furnace exit in a boiler unit. If the analysis is confined to the immediate vicinity of the opening, one can follow the previous example. That is, the opening has the same temperature as the duct wall, unit emissivity, zero heat transfer coefficient, and zero resistance to gas flows, like a porous wall. If the radiant heat transfer between the system and the duct across the opening does substantially affect the results of the system interior, the system analysis should be extended into the entrance region of the duct.

The hatched domain in Fig. 6.1 belongs to the exterior of the enclosed wall (shown by the heavy solid lines) and the elements inside the domain are excluded from the analysis. Gas element 18 may be considered a rectangular solid enclosed by wall elements 23 through 26. Or it could be treated as a rectangular cavity enclosed by wall elements 23 through 26. Either way, the present analysis determines the wall heat flux when the wall emissivity and temperature are specified as the thermal boundary conditions. Or, it is to evaluate the wall temperature with the wall emissivity and heat flux being specified as the thermal boundary conditions. Hence, considering element 18 to be a solid, the program can obtain the heat flux of each wall surface, when this solid temperature is given as the temperatures of the four wall surfaces, elements 23 through 26. The solid temperature must be modified such that the heat flow components into and out of element 18 obtained from the analysis are balanced. The problem of radiative heat transfer to a solid enclosed by a radiant gas is thereby solved. If convective heat transfer is present in addition to the radiant heat transfer between a solid and its surrounding gas, the heat transfer coefficients between the four wall surfaces and their adjacent gas elements must be defined. With the values of these heat transfer coefficients given as the boundary conditions, the corresponding convective heat fluxes are calculated and added to the radiant heat fluxes.

The present program is depicted in Fig. 6.2. The program consists of one main routine and three subroutines, as seen in Fig. 6.3. The title inside each block in the main routine corresponds to that illustrated on the left side of the flow chart in Fig. 4.2. The portion of the RADIAN program for determining the READ values (READC) is essentially identical to the RAT2 program in Section 3.3.5.2. Because the system geometry used in the RAT2 program is rectangular whereas the RADIAN program treats an arbitrary system geometry, some necessary modifications are provided in the latter. The subroutine NXTGAS identifies gas element NG, which is adjacent to wall element NW. It also identifies the index IW, which indicates the position of the wall element relative to the gas element. Subroutine RANDOM determines the uniform random number array

```
1     ************************************************************
2     *                                                          *
3     *                        RADIAN                            *
4     *                                                          *
5     *      RADIATION- AND CONVECTION-HEAT TRANSFER ANALYSIS    *
6     *                   WITHIN AN ENCLOSURE                    *
7     ************************************************************
8   C
9     ************************************************************
10    * (SPC): SPECIFICATION STATEMENT
11    ************************************************************
12  C
13          REAL*8 RAND
14          COMMON /NTG/INDNXT(4,40),IWMAX,NGMAX
15          COMMON /PRT/GP(40),WP(28)
16          DIMENSION INDGW(40),INDNT1(4,20),INDNT2(4,20),AK(40),CP(40),
17         1          TG(40),QG(40),GMF(4,40),GMF1(4,20),GMF2(4,20),
18         2          INDWBC(28),DLW(28),SW(28),TW(28),QW(28),EM(28),H(28),
19         3          RDGG(40,40),RDGW(40,28),RDWG(28,40),RDWW(28,28),
20         4          ASG(40),ASW(28),S(4),ANEWG(40),BNEWG(40),
21         5          ANEWW(28),BNEWW(28)
22  C
23    ************************************************************
24    * (IDATA): FIXED AND INITIAL DATA
25    ************************************************************
26          DATA NGM,NWM/25,20/
27    *------------------------
28    * DATA FOR GAS ELEMENTS
29    *------------------------
30          DATA INDGW/25*1,           15*0/
31          DATA AK/11*0.1,3*0.7,2*0.1,3*0.7,3*0.1,0.7,2*0.1,    15*0./
32          DATA CP/25*1000.,          15*0./,  CP0/1000./
33          DATA TG/25*820.,           15*0./,  TG0/573./
34          DATA QG/11*0.,3*7.E5,2*0.,3*7.E5,3*0.,7.E5,2*0.,     15*0./
35          DATA GMF1/
36         1    0.,-3.,0.,3.,      0.,-3.,0.,3.,      0.,-3.,0.,3.,
37         2    0.,-3.,0.,3.,      0.,-3.,0.,3.,      0.,-3.,0.,3.,
38         3    0.,-3.,0.,3.,      0.,-3.,0.,3.,      0.,-3.,0.,3.,
39         4    0.,-3.,0.,3.,      0.,-3.,0.,3.,      0.,-3.,0.,3.,
40         5    0.,-3.,0.,3.,      0.,-3.,0.,3.,      0.,-3.,0.,3.,
41         6    0.,-3.,0.,3.,      0.,-3.,0.,3.,      0.,-3.,0.,3.,
42         7    0.,-3.,0.,3.,      0.,-3.,0.,3./
43          DATA GMF2/
44         1    0.,-3.,0.,3.,      0.,-3.,0.,3.,      0.,-3.,0.,3.,
45         2    0.,-3.,0.,3.,      0.,-3.,0.,3.,      60*0./
46          DATA INDNT1/
47         1 20,1,-2,-6,       -1,2,-3,-7,       -2,3,-4,-8,      -3,4,-5,-9,
48         2 -4,5,6,-10,       19,-1,-7,-11,     -6,-2,-8,-12,    -7,-3,-9,-13,
49         3 -8,-4,-10,-14,    -9,-5,7,-15,      18,-6,-12,-16,   -11,-7,-13,-17,
50         4 -12,-8,-14,-18,   -13,-9,-15,-19,   -14,-10,8,-20,   17,-11,-17,-21,
51         5 -16,-12,-18,-22,  -17,-13,-19,-23,  -18,-14,-20,-24, -19,-15,9,-25/
52          DATA INDNT2/
53         1 16,-16,-22,15,    -21,-17,-23,14,   -22,-18,-24,13,  -23,-19,-25,12,
54         2 -24,-20,10,11,    60*0/
55          DATA DXG,DYG/1.,1./
56    *------------------------
57    * DATA FOR WALL ELEMENTS
58    *------------------------
59          DATA INDWBC/10*1,5*0,5*1,      8*0/
```

FIG. 6.2. Program of combined radiation and convection heat transfer analysis by the Monte Carlo method

6.2. RADIATIVE HEAT TRANSFER: PROGRAM RADIAN

```
60          DATA DLW/20*1.,              8*0./
61          DATA TW/20*670.,             8*0./
62          DATA QW/20*0.,               8*0./,    EM/20*0.8,        8*0./
63          DATA H/5*0.,5*15.,5*0.,5*15.,  8*0./
64    *-----------------------------
65    * ZERO SETTING OF 'READ' VALUES
66    *-----------------------------
67          DATA RDGG,RDGW,RDWG,RDWW/4624*0./
68    *-----------------------------
69    * INITIAL SETTING OF VARIABLES
70    *-----------------------------
71          NGMAX=NGM
72          NWMAX=NWM
73          IWMAX=4
74          DO 5000 I=1,NWMAX
75             SW(I)=DLW(I)*1.0
76    5000 CONTINUE
77          DO 5010 NG=1,NGMAX
78             DO 5020 IW=1,IWMAX
79                IF(NG.LE.20) THEN
80                   INDNXT(IW,NG)=INDNT1(IW,NG)
81                   GMF(IW,NG)=GMF1(IW,NG)
82                ELSE
83                   NG2=NG-20
84                   INDNXT(IW,NG)=INDNT2(IW,NG2)
85                   GMF(IW,NG)=GMF2(IW,NG2)
86                END IF
87    5020    CONTINUE
88    5010 CONTINUE
89    *------------
90    * CONSTANTS
91    *------------
92          PAI=3.14159
93          SBC=5.6687E-8
94          RAND=5249347.D0
95          VG=DYG*DXG*1.0
96          open ( 6,file='PRN' )
97    *-------------------------------------
98    * INPUT OF CALCULATIONAL CONDITIONS
99    *-------------------------------------
100         write(*,100)
101   100 format(1h0,'input energy particle numbers emitted'/
102        1' from an element (NRAY)')
103         READ(*,*) NRAY
104         write(*,102)
105   102 format(1h0,'for luminous/non-luminous flame,'/
106        1' input 1 or 0')
107         READ(*,*) INDFL
108         IF(INDFL.EQ.0) THEN
109            DO 5030 NG=1,NGMAX
110               AK(NG)=0.2
111   5030    CONTINUE
112         END IF
113         write(*,104)
114   104 format(1h0,'for full/half load, input 1 or 0')
115         READ(*,*) INDFUL
116         IF(INDFUL.EQ.0) THEN
117            DO 5040 NG=1,NGMAX
118               QG(NG)=QG(NG)*0.5
```

FIG. 6.2. (*Continued*)

```
119            DO 5050 IW=1,IWMAX
120               GMF(IW,NG)=GMF(IW,NG)*0.5
121     5050   CONTINUE
122     5040  CONTINUE
123         END IF
124         write(*,106)
125      106 format(1h0,'want to print out ''READ'' values?'/
126        1' yes:1, no:0')
127         READ(*,*) INDRDP
128    C
129    *************************************************************************
130    * (READC): CALCULATION OF "READ" VALUES
131    *************************************************************************
132    C
133    *------------------
134    * FOR GAS ELEMENTS
135    *------------------
136         write(*,110)
137      110 format(1h0,'calculation of ''READ''values'/)
138         DO 5060 NG=1,NGMAX
139          IF(INDGW(NG).EQ.1) THEN
140            DO 5070 INRAY=1,NRAY
141             NGET=NG
142             INDGWC=1
143             INDABS=0
144    * (START POINT)
145             CALL RANDOM(RAN,RAND)
146             X0=(RAN-0.5)*DXG
147             CALL RANDOM(RAN,RAND)
148             Y0=(RAN-0.5)*DYG
149    * (EMITTED DIRECTION)
150             CALL RANDOM(RAN,RAND)
151             ETA=ACOS(1.0-2.0*RAN)
152             CALL RANDOM(RAN,RAND)
153             THTA=2.0*PAI*RAN
154             AL=SIN(ETA)*COS(THTA)
155             AM=COS(ETA)
156             IF(ABS(AL).LT.1.E-10) THEN
157              AL=SIGN(1.E-10,AL)
158             END IF
159             IF(ABS(AM).LT.1.E-10) THEN
160              AM=SIGN(1.E-10,AM)
161             END IF
162    * (ABSORPTION LENGTH)
163             CALL RANDOM(RAN,RAND)
164             RAN1=1.0-RAN
165             IF(RAN1.LT.1.E-6) THEN
166              RAN1=1.E-6
167             END IF
168             XK=-ALOG(RAN1)
169    *
170             XI=X0
171             YI=Y0
172     5080   CONTINUE
173             IF(INDGWC.EQ.1) THEN
174              NGE=NGET
175              S(1)=-(0.5*DXG+XI)/AL
176              S(2)= (0.5*DYG-YI)/AM
177              S(3)= (0.5*DXG-XI)/AL
```

FIG. 6.2. (*Continued*)

6.2. RADIATIVE HEAT TRANSFER: PROGRAM RADIAN

```
178              S(4)=-(0.5*DYG+YI)/AM
179              SMIN=1.E20
180              DO 5090 I=1,IWMAX
181                IF((S(I).GT.1.E-4).AND.(S(I).LT.SMIN)) THEN
182                  SMIN=S(I)
183                  IW=I
184                END IF
185   5090       CONTINUE
186              XK=XK-SMIN*AK(NGE)
187              XE=XI+SMIN*AL
188              YE=YI+SMIN*AM
189              IF(INDNXT(IW,NGE).GT.0) THEN
190                INDGWC=0
191                NWE=INDNXT(IW,NGE)
192              ELSE
193                INDGWC=1
194                NGET=-INDNXT(IW,NGE)
195                IF(IW.EQ.1) THEN
196                  XI=0.5*DXG
197                  YI=YE
198                ELSE IF(IW.EQ.2) THEN
199                  XI=XE
200                  YI=-0.5*DYG
201                ELSE IF(IW.EQ.3) THEN
202                  XI=-0.5*DXG
203                  YI=YE
204                ELSE IF(IW.EQ.4) THEN
205                  XI=XE
206                  YI=0.5*DYG
207                END IF
208              END IF
209              IF(XK.LE.0.) THEN
210                RDGG(NG,NGE)=RDGG(NG,NGE)+1.0
211                INDABS=1
212              END IF
213              ELSE
214                CALL RANDOM(RAN,RAND)
215                IF(RAN.LE.EM(NWE)) THEN
216                  RDGW(NG,NWE)=RDGW(NG,NWE)+1.0
217                  INDABS=1
218                ELSE
219                  CALL RANDOM(RAN,RAND)
220                  ETA=ACOS(SQRT(1.-RAN))
221                  CALL RANDOM(RAN,RAND)
222                  THTA=2.0*PAI*RAN
223                  IF(IW.EQ.1) THEN
224                    AL=COS(ETA)
225                    AM=SIN(ETA)*SIN(THTA)
226                  ELSE IF(IW.EQ.2) THEN
227                    AL=SIN(ETA)*COS(THTA)
228                    AM=-COS(ETA)
229                  ELSE IF(IW.EQ.3) THEN
230                    AL=-COS(ETA)
231                    AM=SIN(ETA)*SIN(THTA)
232                  ELSE IF(IW.EQ.4) THEN
233                    AL=SIN(ETA)*COS(THTA)
234                    AM=COS(ETA)
235                  END IF
236                  IF(ABS(AL).LT.1.E-10) THEN
```

FIG. 6.2. (*Continued*)

```
237                         AL=SIGN(1.E-10,AL)
238                       END IF
239                       IF(ABS(AM).LT.1.E-10) THEN
240                         AM=SIGN(1.E-10,AM)
241                       END IF
242                       INDGWC=1
243                       NGET=NGE
244                       XI=XE
245                       YI=YE
246                     END IF
247                   END IF
248                   IF(INDABS.NE.1) GO TO 5080
249   5070          CONTINUE
250           END IF
251           write(*,120) ng
252    120    format(1h ,'NG=',i3)
253   5060 CONTINUE
254 *--------------------
255 * FOR WALL ELEMENTS
256 *--------------------
257         DO 5100 NW=1,NWMAX
258           DO 5110 INRAY=1,NRAY
259             CALL NXTGAS(NW,NG,IW)
260             NGET=NG
261             INDGWC=1
262             INDABS=0
263 * (STARTING POINT)
264             CALL RANDOM(RAN,RAND)
265             IF(IW.EQ.1) THEN
266               X0=-0.5*DXG
267               Y0=(RAN-0.5)*DYG
268             ELSE IF(IW.EQ.2) THEN
269               X0=(RAN-0.5)*DXG
270               Y0=0.5*DYG
271             ELSE IF(IW.EQ.3) THEN
272               X0=0.5*DXG
273               Y0=(RAN-0.5)*DYG
274             ELSE IF(IW.EQ.4) THEN
275               X0=(RAN-0.5)*DXG
276               Y0=-0.5*DYG
277             END IF
278 * (EMITTED DIRECTION)
279             CALL RANDOM(RAN,RAND)
280             ETA=ACOS(SQRT(1.-RAN))
281             CALL RANDOM(RAN,RAND)
282             THTA=2.0*PAI*RAN
283             IF(IW.EQ.1) THEN
284               AL=COS(ETA)
285               AM=SIN(ETA)*SIN(THTA)
286             ELSE IF(IW.EQ.2) THEN
287               AL=SIN(ETA)*COS(THTA)
288               AM=-COS(ETA)
289             ELSE IF(IW.EQ.3) THEN
290               AL=-COS(ETA)
291               AM=SIN(ETA)*SIN(THTA)
292             ELSE IF(IW.EQ.4) THEN
293               AL=SIN(ETA)*COS(THTA)
294               AM=COS(ETA)
295             END IF
```

FIG. 6.2. (*Continued*)

6.2. RADIATIVE HEAT TRANSFER: PROGRAM RADIAN

```
296            IF(ABS(AL).LT.1.E-10) THEN
297              AL=SIGN(1.E-10,AL)
298            END IF
299            IF(ABS(AM).LT.1.E-10) THEN
300              AM=SIGN(1.E-10,AM)
301            END IF
302 *  (ABSORPTION LENGTH)
303            CALL RANDOM(RAN,RAND)
304            RAN1=1.0-RAN
305            IF(RAN1.LT.1.E-6) THEN
306              RAN1=1.E-6
307            END IF
308            XK=-ALOG(RAN1)
309 *
310            XI=X0
311            YI=Y0
312  5120   CONTINUE
313            IF(INDGWC.EQ.1) THEN
314              NGE=NGET
315              S(1)=-(0.5*DXG+XI)/AL
316              S(2)= (0.5*DYG-YI)/AM
317              S(3)= (0.5*DXG-XI)/AL
318              S(4)=-(0.5*DYG+YI)/AM
319              SMIN=1.E20
320              DO 5130 I=1,IWMAX
321                IF((S(I).GT.1.E-4).AND.(S(I).LT.SMIN)) THEN
322                  SMIN=S(I)
323                  IW=I
324                END IF
325  5130       CONTINUE
326            XK=XK-SMIN*AK(NGE)
327            XE=XI+SMIN*AL
328            YE=YI+SMIN*AM
329            IF(INDNXT(IW,NGE).GT.0) THEN
330              INDGWC=0
331              NWE=INDNXT(IW,NGE)
332            ELSE
333              INDGWC=1
334              NGET=-INDNXT(IW,NGE)
335              IF(IW.EQ.1) THEN
336                XI=0.5*DXG
337                YI=YE
338              ELSE IF(IW.EQ.2) THEN
339                XI=XE
340                YI=-0.5*DYG
341              ELSE IF(IW.EQ.3) THEN
342                XI=-0.5*DXG
343                YI=YE
344              ELSE IF(IW.EQ.4) THEN
345                XI=XE
346                YI=0.5*DYG
347              END IF
348            END IF
349            IF(XK.LE.0) THEN
350              RDWG(NW,NGE)=RDWG(NW,NGE)+1.0
351              INDABS=1
352            END IF
353            ELSE
354              CALL RANDOM(RAN,RAND)
```

FIG. 6.2. (*Continued*)

```
355                    IF(RAN.LE.EM(NWE)) THEN
356                      RDWW(NW,NWE)=RDWW(NW,NWE)+1.0
357                      INDABS=1
358                    ELSE
359                      CALL RANDOM(RAN,RAND)
360                      ETA=ACOS(SQRT(1.-RAN))
361                      CALL RANDOM(RAN,RAND)
362                      THTA=2.0*PAI*RAN
363                      IF(IW.EQ.1) THEN
364                        AL=COS(ETA)
365                        AM=SIN(ETA)*SIN(THTA)
366                      ELSE IF(IW.EQ.2) THEN
367                        AL=SIN(ETA)*COS(THTA)
368                        AM=-COS(ETA)
369                      ELSE IF(IW.EQ.3) THEN
370                        AL=-COS(ETA)
371                        AM=SIN(ETA)*SIN(THTA)
372                      ELSE IF(IW.EQ.4) THEN
373                        AL=SIN(ETA)*COS(THTA)
374                        AM=COS(ETA)
375                      END IF
376                      IF(ABS(AL).LT.1.E-10) THEN
377                        AL=SIGN(1.E-10,AL)
378                      END IF
379                      IF(ABS(AM).LT.1.E-10) THEN
380                        AM=SIGN(1.E-10,AM)
381                      END IF
382                      INDGWC=1
383                      NGET=NGE
384                      XI=XE
385                      YI=YE
386                    END IF
387                  END IF
388                  IF(INDABS.NE.1) GO TO 5120
389    5110        CONTINUE
390               write(*,130) nw
391      130      format(1h ,' NW=',i3)
392    5100  CONTINUE
393  *-------------------------------
394  * NORMALIZATION OF "READ" VALUES
395  *-------------------------------
396          ANRAY=FLOAT(NRAY)
397          DO 5140 I=1,NGMAX
398            IF(INDGW(I).EQ.1) THEN
399              ASG(I)=RDGG(I,I)/ANRAY
400              OUTRAY=ANRAY-RDGG(I,I)
401              DO 5150 J=1,NGMAX
402                IF(I.EQ.J) THEN
403                  RDGG(I,J)=0.0
404                ELSE
405                  RDGG(I,J)=RDGG(I,J)/OUTRAY
406                END IF
407    5150      CONTINUE
408              DO 5160 J=1,NWMAX
409                RDGW(I,J)=RDGW(I,J)/OUTRAY
410    5160      CONTINUE
411            END IF
412    5140  CONTINUE
413          DO 5170 I=1,NWMAX
```

FIG. 6.2. (*Continued*)

6.2. RADIATIVE HEAT TRANSFER: PROGRAM RADIAN

```
414              ASW(I)=RDWW(I,I)/ANRAY
415              OUTRAY=ANRAY-RDWW(I,I)
416              DO 5180 J=1,NGMAX
417                IF(INDGW(J).EQ.1) THEN
418                  RDWG(I,J)=RDWG(I,J)/OUTRAY
419                END IF
420    5180     CONTINUE
421              DO 5190 J=1,NWMAX
422                IF(I.EQ.J) THEN
423                  RDWW(I,J)=0.0
424                ELSE
425                  RDWW(I,J)=RDWW(I,J)/OUTRAY
426                END IF
427    5190     CONTINUE
428    5170   CONTINUE
429            write(*,140)
430    140   format(1h ,'end of "read" calculation')
431  C
432  ************************************************************************
433  * (TEMP): TEMPERATURE CALCULATION
434  ************************************************************************
435  C
436  *-----------------------------------------------------------------------
437  * CALCULATION OF POLYNOMIAL COEFFICIENTS OF ENERGY BALANCE EQUATIONS
438  *-----------------------------------------------------------------------
439            write(*,150)
440    150   format(1h ,'temperature calculation')
441            DO 5200 NG=1,NGMAX
442              IF(INDGW(NG).EQ.1) THEN
443                ANEWG(NG)=4.0*(1.0-ASG(NG))*SBC*AK(NG)*VG
444                BNEWG(NG)=0.0
445                DO 5210 IW=1,IWMAX
446                  NWH=INDNXT(IW,NG)
447                  IF(NWH.GT.0) THEN
448                    BNEWG(NG)=BNEWG(NG)+H(NWH)*SW(NWH)
449                  END IF
450                  GM=GMF(IW,NG)
451                  IF(GM.LT.0.0) THEN
452                    BNEWG(NG)=BNEWG(NG)-GM*CP(NG)
453                  END IF
454    5210       CONTINUE
455              END IF
456    5200   CONTINUE
457            DO 5220 NW=1,NWMAX
458              ANEWW(NW)=(1.0-ASW(NW))*EM(NW)*SBC*SW(NW)
459              BNEWW(NW)=H(NW)*SW(NW)
460    5220   CONTINUE
461  *-----------------------------------------------------------------------
462  * ITERATIONAL CALCULATION OF TEMPERATURE
463  *-----------------------------------------------------------------------
464    5230   CONTINUE
465            ERR=0.0
466            DO 5240 NG=1,NGMAX
467              IF(INDGW(NG).EQ.1) THEN
468                CNEWG=QG(NG)*VG
469                DO 5250 IW=1,IWMAX
470                  NWH=INDNXT(IW,NG)
471                  IF(NWH.GT.0) THEN
472                    CNEWG=CNEWG+H(NWH)*SW(NWH)*TW(NWH)
```

FIG. 6.2. (*Continued*)

```
473                END IF
474                GM=GMF(IW,NG)
475                IF(GM.GT.0) THEN
476                   IUP=-INDNXT(IW,NG)
477                   IF(IUP.GT.0) THEN
478                      CNEWG=CNEWG+GM*CP(IUP)*TG(IUP)
479                   ELSE
480                      CNEWG=CNEWG+GM*CP0*TG0
481                   END IF
482                END IF
483   5250       CONTINUE
484                QRING=0.0
485                DO 5260 NGS=1,NGMAX
486                   IF(INDGW(NGS).EQ.1) THEN
487                      QRING=QRING+RDGG(NGS,NG)*ANEWG(NGS)*TG(NGS)**4
488                   END IF
489   5260       CONTINUE
490                DO 5270 NWS=1,NWMAX
491                   QRING=QRING+RDWG(NWS,NG)*ANEWW(NWS)*TW(NWS)**4
492   5270       CONTINUE
493                CNEWG=-(CNEWG+QRING)
494                TN=TG(NG)
495   5280       CONTINUE
496                DELTAT=((ANEWG(NG)*TN**3+BNEWG(NG))*TN+CNEWG)/
497      1                (4.0*ANEWG(NG)*TN**3+BNEWG(NG))
498                TN=TN-DELTAT
499                ERRN=ABS(DELTAT/TN)
500                IF(ERRN.GE.1.0E-5) GO TO 5280
501                ERRG=ABS((TG(NG)-TN)/TN)
502                IF(ERRG.GT.ERR) THEN
503                   ERR=ERRG
504                END IF
505                TG(NG)=TN
506             END IF
507   5240    CONTINUE
508             DO 5290 NW=1,NWMAX
509                IF(INDWBC(NW).EQ.0) THEN
510                   CALL NXTGAS(NW,NG,IW)
511                   CNEWW=QW(NW)*SW(NW)-H(NW)*TG(NG)*SW(NW)
512                   QRINW=0.0
513                   DO 5300 NGS=1,NGMAX
514                      IF(INDGW(NGS).EQ.1) THEN
515                         QRINW=QRINW+RDGW(NGS,NW)*ANEWG(NGS)*TG(NGS)**4
516                      END IF
517   5300          CONTINUE
518                   DO 5310 NWS=1,NWMAX
519                      QRINW=QRINW+RDWW(NWS,NW)*ANEWW(NWS)*TW(NWS)**4
520   5310          CONTINUE
521                   CNEWW=CNEWW-QRINW
522                   TN=TW(NW)
523   5320       CONTINUE
524                   DELTAT=((ANEWW(NW)*TN**3+BNEWW(NW))*TN+CNEWW)/
525      1                (4.0*ANEWW(NW)*TN**3+BNEWW(NW))
526                   TN=TN-DELTAT
527                   ERRN=ABS(DELTAT/TN)
528                   IF(ERRN.GE.1.0E-5) GO TO 5320
529                   ERRW=ABS((TW(NW)-TN)/TN)
530                   IF(ERRW.GT.ERR) THEN
531                      ERR=ERRW
```

FIG. 6.2. (*Continued*)

6.2. RADIATIVE HEAT TRANSFER: PROGRAM RADIAN

```
532                END IF
533                TW(NW)=TN
534             END IF
535   5290   CONTINUE
536             write(*,160) err
537    160   format(1h ,2x,'ERR=',e12.5)
538          IF(ERR.GE.1.0E-5) GO TO 5230
539          write(*,170)
540    170   format(1h ,'end of temperature calculation')
541 C
542 ************************************************************
543 * (HTFLX): WALL-HEAT FLUX CALCULATION
544 ************************************************************
545 C
546          DO 5330 NW=1,NWMAX
547             IF(INDWBC(NW).EQ.1) THEN
548                CALL NXTGAS(NW,NG,IW)
549                QRINW=0.0
550                DO 5340 NGS=1,NGMAX
551                   IF(INDGW(NGS).EQ.1) THEN
552                      QRINW=QRINW+RDGW(NGS,NW)*ANEWG(NGS)*TG(NGS)**4
553                   END IF
554   5340         CONTINUE
555                DO 5350 NWS=1,NWMAX
556                   QRINW=QRINW+RDWW(NWS,NW)*ANEWW(NWS)*TW(NWS)**4
557   5350         CONTINUE
558                QW(NW)=(QRINW-ANEWW(NW)*TW(NW)**4)/SW(NW)
559       1            +H(NW)*(TG(NG)-TW(NW))
560             END IF
561   5330   CONTINUE
562 C
563 ************************************************************
564 * (OUTPUT): PRINT-OUT OF RESULTS
565 ************************************************************
566 C
567 *---------------------
568 * TITLE AND INPUT DATA
569 *---------------------
570          WRITE(6,200)
571    200   FORMAT(1H ,'*** COMBINED RADIATION CONVECTION HEAT TRANSFER IN FUR
572       1NACE ***'//)
573          WRITE(6,210) NRAY
574    210   FORMAT(1H ,2X,'NUMBER OF ENERGY PARTICLES EMITTED FROM EACH ELEMEN
575       1T=',I7/)
576          AKMAX=0.0
577          DO 5353 I=1,NGMAX
578             IF(AKMAX.LT.AK(I)) THEN
579                AKMAX=AK(I)
580             END IF
581   5353   CONTINUE
582          IF(INDFL.EQ.1) THEN
583             WRITE(6,220) AKMAX
584    220      FORMAT(1H ,2X,'LUMINOUS FLAME (MAXIMUM GAS ABSORPTION COEFF.
585       1    =',F7.3,')'/)
586          ELSE
587             WRITE(6,230) AKMAX
588    230      FORMAT(1H ,2X,'NON-LUMINOUS FLAME (MAXIMUM GAS ABSORPTION
589       1 COEFF. =',F7.3,')'/)
590          END IF
```

FIG. 6.2. (*Continued*)

```
591           TMASSF=0.0
592           DO 5355 I=1,NWMAX
593             CALL NXTGAS(I,NG,IW)
594             GMFIN=GMF(IW,NG)
595             IF(GMFIN.GT.0.0) THEN
596               TMASSF=TMASSF+GMFIN
597             END IF
598   5355  CONTINUE
599           AVHG=0.0
600           NGIN=0
601           DO 5360 I=1,NGMAX
602             IF(INDGW(I).EQ.1) THEN
603               NGIN=NGIN+1
604               AVHG=AVHG+QG(I)
605             END IF
606   5360  CONTINUE
607           AVHG=AVHG/FLOAT(NGIN)
608           IF(INDFUL.EQ.1) THEN
609             WRITE(6,240)
610    240    FORMAT(1H ,2X,'FULL LOAD'/)
611           ELSE
612             WRITE(6,250)
613    250    FORMAT(1H ,2X,'HALF LOAD'/)
614           END IF
615           WRITE(6,260) TMASSF,AVHG
616    260  FORMAT(1H ,2X,'TOTAL INCOMING MASS FLOW=',E13.5,' KG/S/M',10X,
617         1'AVERAGE HEAT LOAD=',E13.5,' W/M**3')
618   *---------------
619   * "READ" VALUES
620   *---------------
621           IF(INDRDP.EQ.1) THEN
622             WRITE(6,300)
623    300    FORMAT(1H0,'ANALYTICAL RESULTS OF "READ" VALUES')
624    265    continue
625             write(*,270)
626    270    format(1h0,'input the number of starting gas element whose "re
627         1ad" you want to print',/,' (to stop, input -1)')
628             read(*,*) ngprnt
629             if(ngprnt.lt.0) go to 5365
630             DO 5370 I=1,NGMAX
631               if(i.eq.ngprnt) then
632                 WRITE(6,310) I
633    310      FORMAT(1H0,2X,'NUMBER OF STARTING GAS ELEMENT=',I3/)
634               DO 5380 NG=1,NGMAX
635                 IF(I.EQ.NG) THEN
636                   GP(NG)=-1.0
637                 ELSE
638                   GP(NG)=RDGG(I,NG)
639                 END IF
640   5380      CONTINUE
641               DO 5390 NW=1,NWMAX
642                 WP(NW)=RDGW(I,NW)
643   5390      CONTINUE
644               CALL PRTDAT
645               end if
646   5370    CONTINUE
647             go to 265
648   5365    continue
649    312    continue
```

FIG. 6.2. (*Continued*)

6.2. RADIATIVE HEAT TRANSFER: PROGRAM RADIAN

```
650             write(*,314)
651     314     format(1h0,'input the number of starting wall element whose "r
652         lead" you want to print',/,' (to stop, input -1)')
653             read(*,*) wgprnt
654             if(wgprnt.lt.0) go to 5405
655             DO 5400 I=1,NWMAX
656               if(i.eq.wgprnt) then
657                 WRITE(6,320) I
658     320         FORMAT(1H0,2X,'NUMBER OF STARTING WALL ELEMENT=',I3/)
659                 DO 5410 NG=1,NGMAX
660                   GP(NG)=RDWG(I,NG)
661     5410        CONTINUE
662                 DO 5420 NW=1,NWMAX
663                   IF(I.EQ.NW) THEN
664                     WP(NW)=-1.0
665                   ELSE
666                     WP(NW)=RDWW(I,NW)
667                   END IF
668     5420        CONTINUE
669                 CALL PRTDAT
670               end if
671     5400    CONTINUE
672             go to 312
673     5405    continue
674     *----------------------
675     * SELF-ABSORPTION RATIO
676     *----------------------
677             WRITE(6,325)
678     325     FORMAT(1H0,'SELF-ABSORPTION RATIO'/)
679             DO 5422 NG=1,NGMAX
680               GP(NG)=ASG(NG)
681     5422    CONTINUE
682             DO 5424 NW=1,NWMAX
683               WP(NW)=ASW(NW)
684     5424    CONTINUE
685             CALL PRTDAT
686             END IF
687     *------------
688     * TEMPERATURE
689     *------------
690             WRITE(6,330)
691     330     FORMAT(1H0,'ANALYTICAL RESULTS OF TEMPERATURE (K)'/)
692             DO 5430 NG=1,NGMAX
693               GP(NG)=TG(NG)
694     5430    CONTINUE
695             DO 5440 NW=1,NWMAX
696               WP(NW)=TW(NW)
697     5440    CONTINUE
698             CALL PRTDAT
699     *----------------
700     * WALL HEAT FLUX
701     *----------------
702             WRITE(6,340)
703     340     FORMAT(1H0,'ANALYTICAL RESULTS OF HEAT FLUXES (W/M**2)'/)
704             DO 5450 NG=1,NGMAX
705               GP(NG)=0.0
706     5450    CONTINUE
707             DO 5460 NW=1,NWMAX
708               WP(NW)=QW(NW)
```

FIG. 6.2. (*Continued*)

```
709    5460 CONTINUE
710         CALL PRTDAT
711         STOP
712         END
713
714
715         SUBROUTINE NXTGAS(NW,NG,IW)
716         COMMON /NTG/INDNXT(4,40),IWMAX,NGMAX
717         NG=1
718         IW=0
719    6000 CONTINUE
720            IW=IW+1
721            IF(IW.GT.IWMAX) THEN
722              IW=1
723              NG=NG+1
724              IF(NG.GT.NGMAX) THEN
725                WRITE(6,140) NW
726    140        FORMAT(1H ,'WALL ELEMENT (NW=',I3,
727      1                ') IS NOT CONNECTED TO ANY GAS ELEMENTS')
728                STOP
729              END IF
730            END IF
731         IF(NW.NE.INDNXT(IW,NG)) GO TO 6000
732         RETURN
733         END
734
735
736         SUBROUTINE RANDOM(RAN,RAND)
737         REAL*8 RAND
738         RAND=DMOD(RAND*131075.0D0,2147483649.0D0)
739         RAN=SNGL(RAND/2147483649.0D0)
740         RETURN
741         END
742
743
744         SUBROUTINE PRTDAT
745         COMMON /PRT/GP(40),WP(28)
746         CHARACTER*1 G,W,B15
747         DATA G/'G'/,W/'W'/,B15/'               '/
748    7000 FORMAT(1H ,7(2X,A1,E12.5))
749    7010 FORMAT(1H ,A15,5(2X,A1,E12.5),A15)
750    7020 FORMAT(1H ,7A15)
751         WRITE(6,7010) B15,(W,WP(I),I=1,5),B15
752         DO 8000 I=1,5
753            WRITE(6,7000) W,WP(21-I),(G,GP(5*(I-1)+J),J=1,5),W,WP(5+I)
754            IF(I.NE.5) THEN
755              WRITE(6,7020) (B15,J=1,7)
756            END IF
757    8000 CONTINUE
758         WRITE(6,7010) B15,(W,WP(16-I),I=1,5),B15
759         RETURN
760         END
```

FIG. 6.2. (*Continued*)

RAN, which is like the one mentioned in Section 3.3.1. The PRTDAT subroutine prints the output data.

Like programs RAT1 and RAT2, the list of variables for the RADIAN program is given at the end of this text. Because the program has many input data, they are entered into the program using the DATA statements. Lines 26 through 67 in Fig. 6.2 are used for the input of data for the combined radiative–convective heat transfer analysis of the gas domain

6.2. RADIATIVE HEAT TRANSFER: PROGRAM RADIAN

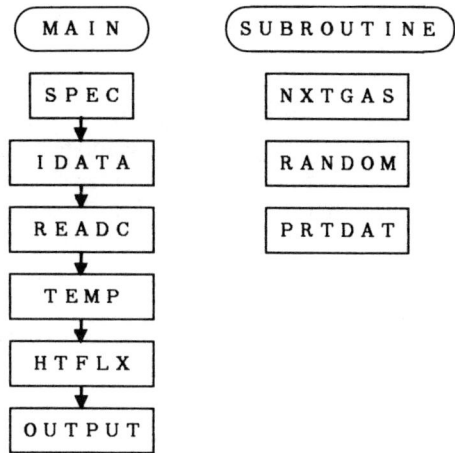

FIG. 6.3. Main routine and subroutines of RADIAN

confined by relatively simple rectangular walls in Fig. 6.4. In this example, the rectangular domain is subdivided into 25 gas elements and 20 wall elements. The numbers in the figure refer to those of the gas and wall elements. The input conditions of this system are given in Fig. 6.5. Wall elements 1 through 5 and 11 through 15 are the walls of a hole opening

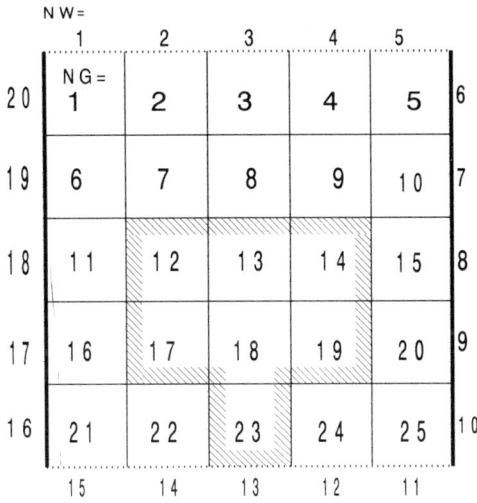

FIG. 6.4. Numbers of gas and wall elements

with an emissivity of 0.8, zero heat transfer coefficient, and zero flow resistance. Their thermal boundary conditions are at a constant temperature of 670 K for wall elements 1 through 5 and insulation for wall elements 11 through 15. A stream of gas flows upward through wall elements 11 through 15 into the system at a velocity of 3m/sec and flows out of the system through wall elements 1 through 5 at the same velocity.

If each wall element is a hole of an identical size, it is treated as a fictitious porous wall at the ambient temperature or the duct temperature at the exit side of the flowing gas. However, if the gas flows in and out of a hole smaller than a wall element, there exists, in reality, a solid at the location of that wall element. Accordingly, the boundary condition must take into account the emissivity and temperature or heat flux of the solid wall together with the convective heat transfer coefficient. In the present example, wall elements 1 through 5 of the exit flow holes have a specified temperature for acting as a cooling wall, whereas wall elements 11 through 15 have zero heat flux for acting as an insulated wall (in the absence of cooling). The other wall elements on both the left and right sides are cooling walls at a temperature of 670 K with an emissivity of 0.8 and a convective heat transfer coefficient of 15 kW/m^2.

The hatched portions in Figs. 6.4 and 6.5 imitate a simplified flame extending upward. Heat is uniformly generated at a volumetric rate of 7×10^5 W/m^3 within the flame. Treating it as a luminant flame, its absorption coefficient of 0.7 m^{-1} is higher than 0.1 m^{-1} for the combustion gas domain outside the flame. In an actual flame, the flame base, in general, produces more heat than the flame tip. Hence, it is necessary, in the actual analysis, to specify a different fraction of heat generation rate for each element inside the flame, by taking into account the distribution of heat generation rates within the flame. The magnitude of absorption coefficients of a nonluminant flame and a combustion gas can be calculated using the method presented in Section 1.3.4. The magnitude of the absorption coefficients of a luminous flame varies significantly with the soot content, but can be evaluated using the method of Kunitomo et al. [27].

The following describes the method for compiling the input data, lines 26 through 67 in the list of Fig. 6.2:

Line 26 specifies the maximum numbers of gas elements NGM and wall elements NWM. The number of gas elements NGM includes not only the gas elements that are the objects of actual numerical computations but also those hatched elements outside the system, as shown in Fig. 6.1. In other words, NGM includes all gas elements that are enclosed in the rectangle. Here, elements 1, 2, 5, 6, 7, 11, 18, and 30 are not the objects of computations but are included in NGM for the case shown in Fig. 6.1. The

6.2. RADIATIVE HEAT TRANSFER: PROGRAM RADIAN

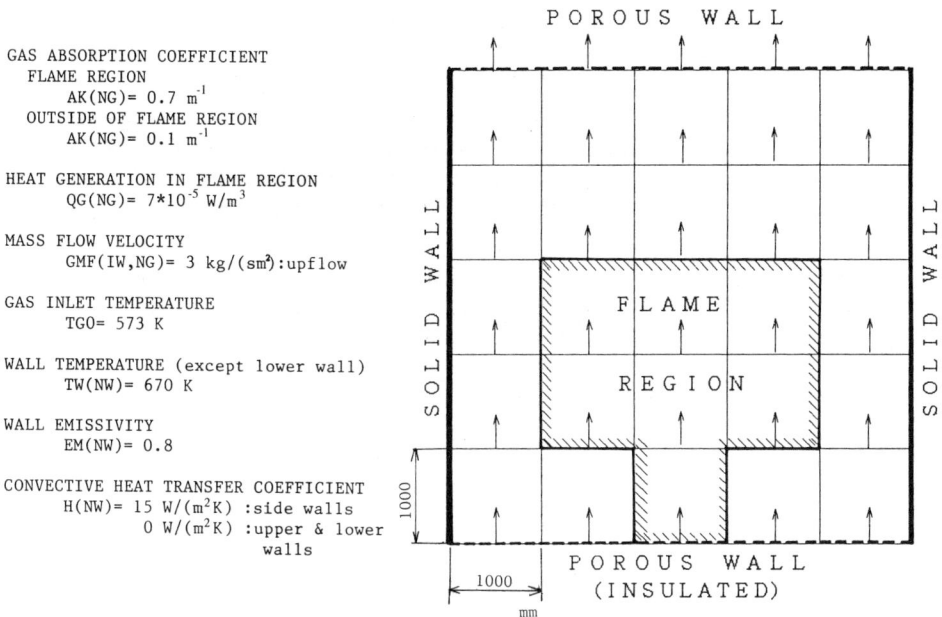

FIG. 6.5. Analytical conditions of RADIAN

gas elements are numbered in order from the top, row by row, and from left toward right, as seen in Fig. 6.1. The wall elements are numbered in a similar manner as the gas elements for convenience in writing the FORMAT statements for the print subroutine, PRTDAT. INDGW in line 30 is the variable to distinguish the gas elements (i.e., computational objects) from those that are not, that is, INDGW = 1 for the former and INDGW = 0 for the latter.

The list in Fig. 6.2 is intended for the system of rectangular geometry shown in Figs. 6.4 and 6.5. Hence, NGM takes the number of gas elements, 25, which are the objects in actual computations. Gas elements 1 through 25 in Figs. 6.4 and 6.5 are identified by INDGW (I) = 1 (I = 1, 25). Since the variable is defined to represent 40 elements, the remaining INDGW (I)'s are identified by 0. In the system of Fig. 6.1, INDGW (I) would be 0 for I = 1, 2, 5, 6, 7, 11, 18, and 30 through 40. AK in line 31 is the absorption coefficient of each gas element (m^{-1}). In Figs. 6.4 and 6.5, AK = 0.7 for gas elements 12, 13, 14, 17, 18, 19, and 23 inside the flame, AK = 0.1 for the other gas elements, and AK = 0 for gas elements 26 through 40, which are not the objects of computations. CP, TG, and QG in lines 32, 33 and 34, respectively, denote the specific heat (K/kg-K), initial temperature (K), and volumetric heat generation rate (W/m^3) for each gas element. Their magnitudes vary depending on the gas elements, such as AK. CPO and TGO are respectively, the constant-pressure specific heat and the temperature of the gas flowing into the system from the bottom, in Figs. 6.4 and 6.5. Lines 35 through 45 express the mass fluxes (kg/s-m^2) of gas flows perpendicularly into and out of the gas elements on the four sides of the rectangular system. The mass flux takes a positive value for an in-flow and negative for an out-flow. For example, 0. −3., 0., and 3. at the beginning of line 36 denote the mass fluxes at the left, top, right, and bottom surfaces of gas element 1 in Figs. 6.4 and 6.5, respectively. It implies that a mass flux of 3 kg/s-m^2 flows into the gas element across the bottom surface and out from the top side at the same mass flux. In the list of Fig. 6.2, the mass flux of GMF1 enters into gas elements 1 through 20 and GMF2 into gas elements 21 through 40. This is because the Fortran system being used can take only 660 characters in one sentence. Hence, the DATA statement is divided into two sentences. The mass flux of the 15 gas elements that are not the objects of computations is set to 0, as in line 45.

INDNT1 and INDNT2 in lines 46 through 54 are used to assign the numbers for the gas and wall elements, respectively, that are adjacent to the four sides of each gas element. A minus number is given to gas elements and a positive number to wall elements. For example, 20, 1, −2, and −6 at the beginning of line 47 denote wall element 20, wall 1, gas 2, and gas 6, respectively which are adjacent to the left, top, right, and bottom sides of gas element 1, respectively. Similar to GMF1 and GMF2, these variables for elements 1 through 20 use INDNT1 for their inputs, and those for elements 21 through 40 use INDNT2. Their value is set to 0 for the elements between 21 and 40 that are not the objects of computations, as seen in line 54. Line 55 enters the width DXG (in meters) and height DYG (also in meters) of gas elements.

INDWBC in line 59 expresses the boundary conditions of each wall element: 1 for the case to determine the wall heat flux when the wall temperature is given, and 0 for vice versa. In the system of Figs. 6.4 and 6.5, wall elements 1 through 10 and 16 through 20 have their temperatures specified as boundary conditions. Wall elements 11 through 15 have their heat fluxes given as the boundary conditions. (In the case of Figs. 6.4 and 6.5, these wall elements are insulated, equivalent to specifying zero heat flux as the boundary conditions.) In line 18, 28 is reserved as the number of wall elements to be arranged. Since only 20 wall elements are used, the remaining 8 elements are designated with INDWBC (I) = 0, as seen at the right end of line 59. The length DLW (m), temperature TW (K), surface heat flux QW (W/m^2), emissivity EM, and convective heat transfer coefficient H (W/m^2-K) of each wall element are described in lines 60 through 63. For a wall element with specified temperatures as the boundary condition, that is, INDWBC (I) = 1, the value of TW is utilized in the boundary condition, whereas any arbitrary value (for example, 0 in line 62) can be given as the input data of the wall heat flux QW. For wall elements with specified heat fluxes as the boundary conditions, that is, INDWBC (I) = 0, the input data of TW are used as the initial values in the wall temperature computations. The statement in line 67 assigns 0 as the initial value of the READ, which corresponds to gas element to gas element, gas element to wall element, wall element to gas element, and wall element to wall element. It need not be changed as the problems change.

All DATA statements regarding the input values are presented in the preceding paragraphs. However, subroutine PRTDAT, following line 744 in Fig. 6.2, varies with each problem. This subroutine arranges the relative positions of the computational results in the printout, as depicted in Fig. 6.6. These analytical results include data for both the gas and wall elements, which are transferred from the main routine to the COMMON array variables, 40 GP (NG)'s for gas elements and 28 GW (NW)'s for wall elements. Lines 748 and 749, (2X, A1, E12.5), describe that the output of each element, irrespective of the gas and wall elements, has 15 digits: 2 empty spaces, one space for G or W to distinguish a gas or wall element, and 12 digits of computational results. FORMAT in lines 749 and 750 contains A15 in order to produce 15 blanks in the output of each element, which is not the object of computations in the rectangular domain. To cope with the output being either a datum or a blank, the FORMAT statements are prepared in lines 748 through 750, depending on the arrangement pattern in a row. Lines 751 through 758 suggest the format of the output data. Consider line 751, for example. Both the left and right ends are blank, sandwiching a character W for wall element and the output data WP (1) through WP (5). The first row in Fig. 6.6(a) (W .27322E-01 W .51613E-01 etc.) is the output of READ values, which are printed according to the WRITE statement.

In the process of executing a program, the computer will ask for four input data on its CRT: NRAY of line 103, INDFL of line 107, INDFUL of line 115, and INDRDP of line 127, in that order. Here, NRAY denotes the number of energy particles used in calculating the READ of each element. Entering 1 into INDFL results in the use of AK values given in the DATA statement of line 31 as the absorption coefficients of each gas element. Should 0 be entered, all AK's will take a value of 0.2 m^{-1}. This is the statement to perform an approximate analysis without changing the DATA statement in line 31, in the case of a nonluminous flame having a small flame absorption coefficient. Should the exact analysis be performed on a

a

*** COMBINED RADIATION CONVECTION HEAT TRANSFER IN FURNACE ***

NUMBER OF ENERGY PARTICLES EMITTED FROM EACH ELEMENT= 50000
LUMINOUS FLAME (MAXIMUM GAS ABSORPTION COEFF. = .700)
FULL LOAD
TOTAL INCOMING MASS FLOW= .15000E+02 KG/S/M   AVERAGE HEAT LOAD= .19600E+06 W/M**3
ANALYTICAL RESULTS OF "READ" VALUES

NUMBER OF STARTING GAS ELEMENT= 8

```
           W .27322E-01  W .51613E-01  W .72422E-01  W .52237E-01  W .26247E-01
W .36050E-01 G .10727E-01  G .16638E-01  G .27064E-01  G .18143E-01  G .95445E-02  W .36136E-01
W .41789E-01 G .11071E-01  G .25817E-01  G -.10000E+01 G .25710E-01  G .11630E-01  W .42542E-01
W .28225E-01 G .65780E-02  G .67908E-01  G .12057E+00  G .68187E-01  G .63845E-02  W .27000E-01
W .10684E-01 G .23216E-02  G .18573E-01  G .20959E-01  G .19046E-01  G .28161E-02  W .10447E-01
W .47078E-02 G .11823E-02  G .14618E-02  G .73733E-02  G .14833E-02  G .11608E-02  W .50517E-02
             W .45788E-02  W .60190E-02  W .35039E-02  W .58256E-02  W .52452E-02
```

b

NUMBER OF STARTING WALL ELEMENT= 18

```
           W .23198E-01  W .40966E-01  W .37220E-01  W .25541E-01  W .17488E-01
W .28446E-02 G .67509E-02  G .13282E-01  G .11238E-01  G .80731E-02  G .68912E-02  W .28646E-01
W .24440E-02 G .25561E-01  G .24139E-01  G .12861E-01  G .71716E-02  G .46075E-02  W .19732E-01
W -.10000E+01 G .94533E-01  G .17212E+00  G .46836E-01  G .16166E-01  G .17028E-02  W .72718E-02
W .19231E-02 G .25561E-01  G .11418E+00  G .35678E-01  G .13862E-01  G .13021E-02  W .59296E-02
W .22236E-02 G .65105E-02  G .10677E-01  G .26543E-01  G .18831E-02  G .12420E-02  W .56892E-02
             W .22176E-01  W .40646E-01  W .16507E-01  W .57894E-02  W .43871E-02
```

Fig. 6.6. Output of RADIAN. READ values correspond to the emission from (a) a gas element and (b) a wall element

6.2. RADIATIVE HEAT TRANSFER: PROGRAM RADIAN

nonluminous flame, the absorption coefficient of the flame must enter the DATA statement in line 31, deleting lines 104 through 112. Entering 1 into the input variable INDFUL results in the use of specified values of both the heat generation rate QG of each gas element in line 34 and the incoming mass fluxes GMF1 and GMF2 in lines 35 through 45. If 0 is entered, an approximate analysis of a 50% load will be performed using one-half of their specified values. The exact analysis requires the distributions of both the heat generation rate inside a flame corresponding to a 50% load and the mass fluxes in lines 34 through 45, deleting lines 114 through 123. Setting the input variable INDRDP to 1 results in a printout of the READ values; INDRDP = 0 results in no printout. Because there are a tremendous number of READ values, the program is written to print out only those of necessary elements. To designate those elements whose READ values need to be printed, enter the element number into "ngprnt" in line 628 in the case of gas elements, or "wgprnt" in line 653 for wall elements. As long as no negative numbers are entered for "ngprnt" and "wgprnt," those entries repeatedly request the number of those elements whose READ values should be printed.

The printout of READ values can be stopped, simply by entering -1.

Figure 6.6 presents the results obtained from the list in Fig. 6.2 using the system of Figs. 6.4 and 6.5 as the input data. The number of energy particles emitted from each element is 50,000. INDFL, INDFUL, and INDRDP all take a value of 1. TOTAL INCOMING MASS FLOW denotes the total mass flow rate of the gas entering or leaving the system and AVERAGE HEAT LOAD signifies the average heat generation rate of the entire system including both the flame and no-flame portions. Figures 6.6(a) and (b) present the READ values of all elements in the system, with the gas and wall elements enclosed in small rectangles as the respective emitting unit. The READ value of the emitting element is given as $-.10000\ E + 01$. It is intended for identifying the emitting element and the value is meaningless. Figure 6.6(a) corresponds to the case where gas element 8 in Fig. 6.4 is the emitting element. It is seen in Fig. 6.6(a) that gas element 3, located immediately above the emitting element, absorbs only 2.7% of all radiant energy emitted from gas element 8. In comparison, gas element 13, which is located immediately below the emitting element, absorbs 12%. This is because element 3 has a low absorption coefficient of $0.1\ m^{-1}$, but element 8, located within the flame, has a higher value of $0.7\ m^{-1}$.

Figure 6.6(b) presents a distribution of the READ values for wall element 18 in Fig. 6.4, which acts as the emitting element. The results are somewhat different from those of RAT2 in Fig. 3.27 because of a difference in the distribution of the wall emissivities. But qualitatively, both results exhibit a similar trend due to the influence of a flame with a higher absorptivity at the central part of the system. For a symmetrical system, the radiant energy emitted from a gas element on the symmetrical axis, as in Fig. 6.6(a), should be symmetrically absorbed by both the left and right sides. In Fig. 6.6(a), the READ values of any symmetrical pairs agree up to two digits. Figure 6.7 presents the distribution of self-absorption ratios for both the gas and wall elements. It is observed that the values are higher inside the flame, which has a higher absorption coefficient.

Figure 6.8 depicts the temperature distribution inside the system, and isotherms are plotted in Fig. 6.9. The gas enters the system from the bottom surface and uniformly exits from the top surface. It is heated midway by the heat generated in the flame. Hence, the gas temperature rapidly rises as it travels upward through the flame. As it exists from the flame tip, the gas stream is gradually cooled down due to its radiation to the wall surfaces, causing a slight decrease in the gas temperature. The gas streams that flow upward around both sides of the flame are not subject to internal heat generation but receive radiation from the flame, resulting in a gradual increase of the gas temperature. Both side walls and the top wall have their temperatures specified at 670 K by the boundary conditions. Meanwhile, the bottom wall is adiabatic, with zero net heat flux as the boundary condition. Its temperature distribution is determined so that the radiative and convective heat fluxes balance to yield zero net heat flux. The results indicate that the bottom wall temperature is maximum at the center (880 K) where radiation from the flame is highest, diminishing toward the sides.

Figure 6.10 illustrates the distribution of net wall heat fluxes. The bottom wall has zero heat flux, whereas the center of the top wall has the highest heat flux due to its exposure to the high-temperature portion of the flame.

6.3. Radiative Heat Transfer between Surfaces Separated by Nonparticipating Gas: Program RADIANW

This section treats radiative heat transfer between multiple surfaces separated by nonparticipating gas or vacuum space. The situation is abundant, for example, inside an artificial satellite suspended in a vacuum space, or an electrical stove placed in a room. The system need not be a vacuum as long as the medium between the radiating surfaces does not absorb radiative energy or contains a sufficiently low concentration of solid particles that scatter radiation. Carbon dioxide and water vapor are typical examples of radiant-energy-absorbing media.

SELF-ABSORPTION RATIO

		W .16480E-01	W .15200E-01	W .39200E-02	W .23400E-02	W .32800E-02	W .15740E-01	W .16640E-01
W	.40800E-02	G .79400E-01	G .74700E-01	G .74600E-01	G .73800E-01	G .76340E-01	W .46200E-02
W	.16200E-02	G .74920E-01	G .70960E-01	G .69620E-01	G .70800E-01	G .74900E-01	W .20000E-02
W	.28200E-02	G .73100E-01	G .34650E+00	G .34234E+00	G .34810E+00	G .74860E-01	W .32200E-02
W	.14840E-01	G .73040E-01	G .34930E+00	G .34442E+00	G .34790E+00	G .73420E-01	W .14160E-01
		G .76000E-01	G .76080E-01	G .36352E+00	G .73500E-01	G .78500E-01	
		W .15160E-01	W .27600E-02	W .78000E-03	W .31400E-02	W .15900E-01	

Fig. 6.7. Output of RADIAN for self-absorption ratio

ANALYTICAL RESULTS OF TEMPERATURE (K)

		W .67000E+03	W .67000E+03	W .67000E+03	W .67000E+03	W .67000E+03	W .67000E+03
W	.67000E+03	G .58665E+03	G .10241E+04	G .11898E+04	G .10236E+04	G .58682E+03	W .67000E+03
W	.67000E+03	G .58399E+03	G .10282E+04	G .12000E+04	G .10277E+04	G .58406E+03	W .67000E+03
W	.67000E+03	G .58109E+03	G .10315E+04	G .12095E+04	G .10312E+04	G .58118E+03	W .67000E+03
W	.67000E+03	G .57806E+03	G .81523E+03	G .10274E+04	G .81484E+03	G .57811E+03	W .67000E+03
W	.67000E+03	G .57542E+03	G .57557E+03	G .80987E+03	G .57545E+03	G .57546E+03	W .67000E+03
		W .79538E+03	W .83087E+03	W .84362E+03	W .82924E+03	W .79143E+03	

Fig. 6.8. Output of RADIAN for temperature distribution

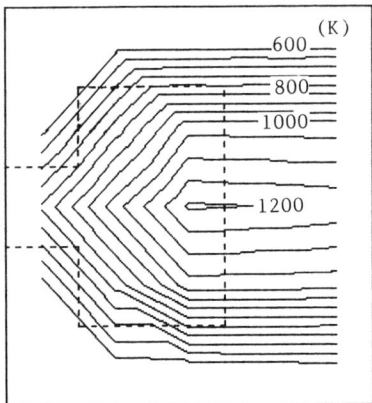

FIG. 6.9. Temperature profile

The problem is a special case or radiant heat transfer among surfaces separated by a radiant gas having zero absorption coefficient. Its analysis can be achieved simply by setting AK (I) = 0 in the RAT2 and RADIAN programs. However, to save computer time for the evaluation of READ values and to reduce the memory space required for computations, the portion dealing with radiant heat exchange with the gas elements is deleted from the RADIAN program. The resulting program, which is depicted in Fig. 6.11, is named RADIANW. Its DATA statements are identical to the corresponding part in the RADIAN. The listing in Fig. 6.11 is applicable to the systems that have a gray solid body B in a space enclosed by gray walls with a protruding part A on the left side, as shown in Fig. 6.12. Both the left and right walls are adiabatic, and radiative energy is transmitted from the lower wall at 1000 K to the upper one at 500 K. The solid body in the middle receives radiation from below and radiates the same amount of energy to the wall above, thus maintaining an overall energy balance. The program is similar to RADIAN, except for the matter concerning the gas. It treats convective heat transfer between the gas and the solid, heat generation in the gas, and enthalpy transport induced by gas flows, and thus can determine the temperature distribution in the gas. Because the DATA statements in the list in Fig. 6.11 assume the system to be a vacuum, the gas specific heat CP(I), initial value TG(I) of gas temperature TGO, heat generation rate in gas QG(I), flow rates GMF1(I) and GMF2(I), and heat transfer coefficient between the gas and the solid walls H(I) are all set to zero. The wall emissivity EM(I) is 0.5. This program is essentially the RADIAN program with the portion of gas

ANALYTICAL RESULTS OF HEAT FLUXES (W/M**2)

		W .15241E+05	W .25039E+05	W .31925E+05	W .24506E+05	W .14577E+05		
W	.13832E+05	G .00000E+00	G .00000E+00	G .00000E+00	G .00000E+00	G .00000E+00	W	.14831E+05
W	.17228E+05	G .00000E+00	G .00000E+00	G .00000E+00	G .00000E+00	G .00000E+00	W	.17806E+05
W	.18315E+05	G .00000E+00	G .00000E+00	G .00000E+00	G .00000E+00	G .00000E+00	W	.18259E+05
W	.15085E+05	G .00000E+00	G .00000E+00	G .00000E+00	G .00000E+00	G .00000E+00	W	.14782E+05
W	.10237E+05	G .00000E+00	G .00000E+00	G .00000E+00	G .00000E+00	G .00000E+00	W	.10258E+05
		W .00000E+00	W .00000E+00	W .00000E+00	W .00000E+00	W .00000E+00		

FIG. 6.10. Output of RADIAN for wall heat flux

```
1     ************************************************************************
2     *                                                                      *
3     *                           RADIANW                                    *
4     *                                                                      *
5     *        RADIATION- AND CONVECTION-HEAT TRANSFER CALCULATION           *
6     *             WITHIN AN ENCLOSURE (NON-PARTICIPATING GAS)              *
7     ************************************************************************
8     C
9     ************************************************************************
10    * (SPC): SPECIFICATION STATEMENT
11    ************************************************************************
12    C
13          REAL*8 RAND
14          COMMON /NTG/INDNXT(4,40),IWMAX,NGMAX
15          COMMON /PRT/GP(40),WP(28)
16          DIMENSION INDGW(40),INDNT1(4,20),INDNT2(4,20),CP(40),
17         1          TG(40),QG(40),GMF(4,40),GMF1(4,20),GMF2(4,20),
18         2          INDWBC(28),DLW(28),SW(28),TW(28),QW(28),EM(28),H(28),
19         3          RDWW(28,28),
20         4          ASW (28), S (4), BNEWG (40)
21         5          ANEWW(28),BNEWW(28)
22    C
23    ************************************************************************
24    * (IDATA): FIXED AND INITIAL DATA
25    ************************************************************************
26          DATA NGM,NWM/25,28/
27    *-------------------------
28    *  DATA FOR GAS ELEMENTS
29    *-------------------------
30          DATA INDGW/10*1,0,1,2*0,11*1,            15*0/
31          DATA CP/40*0./,    CP0/0./
32          DATA TG/40*0./,    TG0/0./
33          DATA QG/40*0./
34          DATA GMF1/80*0./
35          DATA GMF2/80*0./
36          DATA INDNT1/
37         1  19,1,-2,-6,       -1,2,-3,-7,       -2,3,-4,-8,       -3,4,-5,-9,
38         2  -4,5,6,-10,       18,-1,-7,22,      -6,-2,-8,-12,     -7,-3,-9,23,
39         3  -8,-4,-10,24,     -9,-5,7,-15,      0,0,0,0,          21,-7,28,-17,
40         4  0,0,0,0,          0,0,0,0,          25,-10,8,-20,     17,20,-17,-21,
41         5  -16,-12,-18,-22,  -17,27,-19,-23,   -18,26,-20,-24,   -19,-15,9,-25/
42          DATA INDNT2/
43         1  16,-16,-22,15,    -21,-17,-23,14,   -22,-18,-24,13,   -23,-19,-25,12,
44         2  -24,-20,10,11,    60*0/
45          DATA DXG,DYG/1.,1./
46    *-------------------------
47    * DATA FOR WALL ELEMENTS
48    *-------------------------
49          DATA INDWBC/5*1,5*0,5*1,7*0,6*1/
50          DATA DLW/28*1./
51          DATA TW/5*500.,5*750.,5*1000.,13*750./
52          DATA QW/28*0./,    EM/28*0.5/
53          DATA H/28*0./
54    *-------------------------------
55    * ZERO SETTING OF 'READ' VALUES
56    *-------------------------------
57          DATA RDWW/784*0./
58    *-------------------------------
59    * INITIAL SETTING OF VARIABLES
```

FIG. 6.11. Program of combined radiation and convection heat transfer analysis within enclosures containing nonparticipating gas

6.3. RADIATIVE HEAT TRANSFER: PROGRAM RADIANW

```
 60   *----------------------------
 61         NGMAX=NGM
 62         NWMAX=NWM
 63         IWMAX=4
 64         DO 5000 I=1,NWMAX
 65           SW(I)=DLW(I)*1.0
 66   5000 CONTINUE
 67         DO 5010 NG=1,NGMAX
 68           DO 5020 IW=1,IWMAX
 69             IF(NG.LE.20) THEN
 70               INDNXT(IW,NG)=INDNT1(IW,NG)
 71               GMF(IW,NG)=GMF1(IW,NG)
 72             ELSE
 73               NG2=NG-20
 74               INDNXT(IW,NG)=INDNT2(IW,NG2)
 75               GMF(IW,NG)=GMF2(IW,NG2)
 76             END IF
 77   5020    CONTINUE
 78   5010 CONTINUE
 79   *------------
 80   * CONSTANTS
 81   *------------
 82         PAI=3.14159
 83         SBC=5.6687E-8
 84         RAND=5249347.D0
 85         VG=DYG*DXG*1.0
 86         open(6,file='PRN')
 87   *------------------------------------
 88   * INPUT OF CALCULATIONAL CONDITIONS
 89   *------------------------------------
 90         write(*,100)
 91   100 format(1h0,'input energy particle numbers emitted'/
 92       1' from an element (NRAY)')
 93         READ(*,*) NRAY
 94         write(*,106)
 95   106 format(1h0,'want to print out ''READ'' values?'/
 96       1' yes:1, no:0')
 97         READ(*,*) INDRDP
 98   C
 99   *************************************************************
100   * (READC): CALCULATION OF "READ" VALUES
101   *************************************************************
102   C
103   *--------------------
104   * FOR WALL ELEMENTS
105   *--------------------
106         DO 5100 NW=1,NWMAX
107           DO 5110 INRAY=1,NRAY
108             CALL NXTGAS(NW,NG,IW)
109             NGET=NG
110             INDGWC=1
111             INDABS=0
112   * (STARTING POINT)
113             CALL RANDOM(RAN,RAND)
114             IF(IW.EQ.1) THEN
115               X0=-0.5*DXG
116               Y0=(RAN-0.5)*DYG
117             ELSE IF(IW.EQ.2) THEN
118               X0=(RAN-0.5)*DXG
```

FIG. 6.11. (*Continued*)

```
119            Y0=0.5*DYG
120            ELSE IF(IW.EQ.3) THEN
121            X0=0.5*DXG
122            Y0=(RAN-0.5)*DYG
123            ELSE IF(IW.EQ.4) THEN
124            X0=(RAN-0.5)*DXG
125            Y0=-0.5*DYG
126            END IF
127  * (EMITTED DIRECTION)
128            CALL RANDOM(RAN,RAND)
129            ETA=ACOS(SQRT(1.-RAN))
130            CALL RANDOM(RAN,RAND)
131            THTA=2.0*PAI*RAN
132            IF(IW.EQ.1) THEN
133            AL=COS(ETA)
134            AM=SIN(ETA)*SIN(THTA)
135            ELSE IF(IW.EQ.2) THEN
136            AL=SIN(ETA)*COS(THTA)
137            AM=-COS(ETA)
138            ELSE IF(IW.EQ.3) THEN
139            AL=-COS(ETA)
140            AM=SIN(ETA)*SIN(THTA)
141            ELSE IF(IW.EQ.4) THEN
142            AL=SIN(ETA)*COS(THTA)
143            AM=COS(ETA)
144            END IF
145            IF(ABS(AL).LT.1.E-10) THEN
146            AL=SIGN(1.E-10,AL)
147            END IF
148            IF(ABS(AM).LT.1.E-10) THEN
149            AM=SIGN(1.E-10,AM)
150            END IF
151  *
152            XI=X0
153            YI=Y0
154     5120   CONTINUE
155              IF(INDGWC.EQ.1) THEN
156                NGE=NGET
157                S(1)=-(0.5*DXG+XI)/AL
158                S(2)= (0.5*DYG-YI)/AM
159                S(3)= (0.5*DXG-XI)/AL
160                S(4)=-(0.5*DYG+YI)/AM
161                SMIN=1.E20
162                DO 5130 I=1,IWMAX
163                  IF((S(I).GT.1.E-4).AND.(S(I).LT.SMIN)) THEN
164                    SMIN=S(I)
165                    IW=I
166                  END IF
167     5130     CONTINUE
168              XE=XI+SMIN*AL
169              YE=YI+SMIN*AM
170              IF(INDNXT(IW,NGE).GT.0) THEN
171                INDGWC=0
172                NWE=INDNXT(IW,NGE)
173              ELSE
174                INDGWC=1
175                NGET=-INDNXT(IW,NGE)
176                IF(IW.EQ.1) THEN
177                  XI=0.5*DXG
```

FIG. 6.11. (*Continued*)

6.3. RADIATIVE HEAT TRANSFER: PROGRAM RADIANW 137

```
178                   YI=YE
179               ELSE IF(IW.EQ.2) THEN
180                   XI=XE
181                   YI=-0.5*DYG
182               ELSE IF(IW.EQ.3) THEN
183                   XI=-0.5*DXG
184                   YI=YE
185               ELSE IF(IW.EQ.4) THEN
186                   XI=XE
187                   YI=0.5*DYG
188               END IF
189             END IF
190           ELSE
191             CALL RANDOM(RAN,RAND)
192             IF(RAN.LE.EM(NWE)) THEN
193               RDWW(NW,NWE)=RDWW(NW,NWE)+1.0
194               INDABS=1
195             ELSE
196               CALL RANDOM(RAN,RAND)
197               ETA=ACOS(SQRT(1.-RAN))
198               CALL RANDOM(RAN,RAND)
199               THTA=2.0*PAI*RAN
200               IF(IW.EQ.1) THEN
201                   AL=COS(ETA)
202                   AM=SIN(ETA)*SIN(THTA)
203               ELSE IF(IW.EQ.2) THEN
204                   AL=SIN(ETA)*COS(THTA)
205                   AM=-COS(ETA)
206               ELSE IF(IW.EQ.3) THEN
207                   AL=-COS(ETA)
208                   AM=SIN(ETA)*SIN(THTA)
209               ELSE IF(IW.EQ.4) THEN
210                   AL=SIN(ETA)*COS(THTA)
211                   AM=COS(ETA)
212               END IF
213               IF(ABS(AL).LT.1.E-10) THEN
214                   AL=SIGN(1.E-10,AL)
215               END IF
216               IF(ABS(AM).LT.1.E-10) THEN
217                   AM=SIGN(1.E-10,AM)
218               END IF
219               INDGWC=1
220               NGET=NGE
221               XI=XE
222               YI=YE
223             END IF
224           END IF
225           IF(INDABS.NE.1) GO TO 5120
226   5110  CONTINUE
227         write(*,130) nw
228    130  format(1h ,' NW=',i3)
229   5100 CONTINUE
230  *------------------------------
231  * NORMALIZATION OF "READ" VALUES
232  *------------------------------
233         ANRAY=FLOAT(NRAY)
234         DO 5170 I=1,NWMAX
235           ASW(I)=RDWW(I,I)/ANRAY
236           OUTRAY=ANRAY-RDWW(I,I)
```

FIG. 6.11. (*Continued*)

```
237              DO 5190 J=1,NWMAX
238                IF(I.EQ.J) THEN
239                  RDWW(I,J)=0.0
240                ELSE
241                  RDWW(I,J)=RDWW(I,J)/OUTRAY
242                END IF
243  5190      CONTINUE
244  5170    CONTINUE
245         write(*,140)
246    140 format(1h ,'end of "read" calculation')
247 C
248 ************************************************************************
249 * (TEMP): TEMPERATURE CALCULATION
250 ************************************************************************
251 C
252 *-----------------------------------------------------------------------
253 * CALCULATION OF POLYNOMIAL COEFFICIENTS OF ENERGY BALANCE EQUATIONS
254 *-----------------------------------------------------------------------
255         write(*,150)
256    150 format(1h ,'temperature calculation')
257         DO 5200 NG=1,NGMAX
258           IF(INDGW(NG).EQ.1) THEN
259             BNEWG(NG)=0.0
260             DO 5210 IW=1,IWMAX
261               NWH=INDNXT(IW,NG)
262               IF(NWH.GT.0) THEN
263                 BNEWG(NG)=BNEWG(NG)+H(NWH)*SW(NWH)
264               END IF
265               GM=GMF(IW,NG)
266               IF(GM.LT.0.0) THEN
267                 BNEWG(NG)=BNEWG(NG)-GM*CP(NG)
268               END IF
269  5210      CONTINUE
270           END IF
271  5200    CONTINUE
272         DO 5220 NW=1,NWMAX
273           ANEWW(NW)=(1.0-ASW(NW))*EM(NW)*SBC*SW(NW)
274           BNEWW(NW)=H(NW)*SW(NW)
275  5220    CONTINUE
276 *---------------------------
277 * CALCULATION OF TEMPERATURE
278 *---------------------------
279  5230    CONTINUE
280         ERR=0.0
281         DO 5240 NG=1,NGMAX
282           IF(INDGW(NG).EQ.1) THEN
283             CNEWG=QG(NG)*VG
284             DO 5250 IW=1,IWMAX
285               NWH=INDNXT(IW,NG)
286               IF(NWH.GT.0) THEN
287                 CNEWG=CNEWG+H(NWH)*SW(NWH)*TW(NWH)
288               END IF
289               GM=GMF(IW,NG)
290               IF(GM.GT.0) THEN
291                 IUP=-INDNXT(IW,NG)
292                 IF(IUP.GT.0) THEN
293                   CNEWG=CNEWG+GM*CP(IUP)*TG(IUP)
294                 ELSE
295                   CNEWG=CNEWG+GM*CP0*TG0
```

FIG. 6.11. (*Continued*)

6.3. RADIATIVE HEAT TRANSFER: PROGRAM RADIANW

```
296                   END IF
297                END IF
298     5250     CONTINUE
299              IF(BNEWG(NG).NE.0.) THEN
300                 TN=CNEWG/BNEWG(NG)
301                 ERRG=ABS((TG(NG)-TN)/TN)
302              ELSE
303                 TN=0.
304                 ERRG=0.
305              END IF
306              IF(ERRG.GT.ERR) THEN
307                 ERR=ERRG
308              END IF
309              TG(NG)=TN
310           END IF
311     5240  CONTINUE
312           DO 5290 NW=1,NWMAX
313              IF(INDWBC(NW).EQ.0) THEN
314                 CALL NXTGAS(NW,NG,IW)
315                 CNEWW=QW(NW)*SW(NW)-H(NW)*TG(NG)*SW(NW)
316                 QRINW=0.0
317                 DO 5310 NWS=1,NWMAX
318                    QRINW=QRINW+RDWW(NWS,NW)*ANEWW(NWS)*TW(NWS)**4
319     5310        CONTINUE
320                 CNEWW=CNEWW-QRINW
321                 TN=TW(NW)
322     5320        CONTINUE
323                    DELTAT=((ANEWW(NW)*TN**3+BNEWW(NW))*TN+CNEWW)/
324        1                 (4.0*ANEWW(NW)*TN**3+BNEWW(NW))
325                    TN=TN-DELTAT
326                    ERRN=ABS(DELTAT/TN)
327                 IF(ERRN.GE.1.0E-5) GO TO 5320
328                 ERRW=ABS((TW(NW)-TN)/TN)
329                 IF(ERRW.GT.ERR) THEN
330                    ERR=ERRW
331                 END IF
332                 TW(NW)=TN
333              END IF
334     5290  CONTINUE
335           write(*,160) err
336      160  format(1h ,2x,'ERR=',e12.5)
337           IF(ERR.GE.1.0E-5) GO TO 5230
338    *------------------------------------------------------
339    * CALCULATION OF TEMP. OF ISOLATED SOLID BODY
340    *   (This part should be rewritten for each solid body)
341    *------------------------------------------------------
342           ANEWWT=0.0
343           BNEWWT=0.0
344           CNEWWT=0.0
345           DO 5322 NW=23,28
346              CALL NXTGAS(NW,NG,IW)
347              CNEWWT=CNEWWT+QW(NW)*SW(NW)-H(NW)*TG(NG)*SW(NW)
348              QRINW=0.0
349              DO 5324 NWS=1,NWMAX
350                 QRINW=QRINW+RDWW(NWS,NW)*ANEWW(NWS)*TW(NWS)**4
351     5324     CONTINUE
352              CNEWWT=CNEWWT-QRINW
353              ANEWWT=ANEWWT+ANEWW(NW)
354              BNEWWT=BNEWWT+BNEWW(NW)
```

FIG. 6.11. (*Continued*)

```
355       5322 CONTINUE
356            TN=TW(23)
357            DELTAT=((ANEWWT*TN**3+BNEWWT)*TN+CNEWWT)/
358         1         (4.0*ANEWWT*TN**3+BNEWWT)
359            TN=TN-DELTAT
360            ERRN=ABS(DELTAT/TN)
361            DO 5326 NW=23,28
362               TW(NW)=TN
363       5326 CONTINUE
364            write(*,165) errn
365        165 format(1h0,' ERRN=',e12.5)
366            IF(ERRN.GT.1.0E-5) GO TO 5230
367            write(*,170)
368        170 format(1h ,'end of temperature calculation')
369  C
370  **********************************************************************
371  * (HTFLX): WALL-HEAT FLUX CALCULATION
372  **********************************************************************
373  C
374            DO 5330 NW=1,NWMAX
375              IF(INDWBC(NW).EQ.1) THEN
376                CALL NXTGAS(NW,NG,IW)
377                QRINW=0.0
378                DO 5350 NWS=1,NWMAX
379                  QRINW=QRINW+RDWW(NWS,NW)*ANEWW(NWS)*TW(NWS)**4
380       5350     CONTINUE
381                QW(NW)=(QRINW-ANEWW(NW)*TW(NW)**4)/SW(NW)
382         1             +H(NW)*(TG(NG)-TW(NW))
383              END IF
384       5330 CONTINUE
385  C
386  **********************************************************************
387  * (OUTPUT): PRINT-OUT OF RESULTS
388  **********************************************************************
389  C
390  *----------------------
391  * TITLE AND INPUT DATA
392  *----------------------
393            WRITE(6,200)
394        200 FORMAT(1H ,'*** COMBINED RADIATION CONVECTION HEAT TRANSFER IN FUR
395           1NACE ***'/'            (NON-PARTICIPATING GAS ENVIRONMENT)'//)
396            WRITE(6,210) NRAY
397        210 FORMAT(1H ,2X,'NUMBER OF ENERGY PARTICLES EMITTED FROM EACH ELEMEN
398           1T=',I7/)
399            TMASSF=0.0
400            DO 5355 I=1,NWMAX
401              CALL NXTGAS(I,NG,IW)
402              GMFIN=GMF(IW,NG)
403              IF(GMFIN.GT.0.0) THEN
404                TMASSF=TMASSF+GMFIN
405              END IF
406       5355 CONTINUE
407            AVHG=0.0
408            NGIN=0
409            DO 5360 I=1,NGMAX
410              IF(INDGW(I).EQ.1) THEN
411                NGIN=NGIN+1
412                AVHG=AVHG+QG(I)
413              END IF
```

FIG. 6.11. (*Continued*)

6.3. RADIATIVE HEAT TRANSFER: PROGRAM RADIANW

```
414   5360 CONTINUE
415        AVHG=AVHG/FLOAT(NGIN)
416        WRITE(6,260) TMASSF,AVHG
417    260 FORMAT(1H ,2X,'TOTAL INCOMING MASS FLOW=',E13.5,' KG/S/M',10X,
418       1'AVERAGE HEAT LOAD=',E13.5,' W/M**3')
419 *---------------
420 * "READ" VALUES
421 *---------------
422        IF(INDRDP.EQ.1) THEN
423          WRITE(6,300)
424    300   FORMAT(1H0,'ANALYTICAL RESULTS OF "READ" VALUES')
425    312   continue
426            write(*,314)
427    314    format(1h0,'input the number of emitting wall element whose "r
428       1ead" you want to print',/,' (to stop, input -1)')
429            read(*,*) wgprnt
430            if(wgprnt.lt.0) go to 5405
431            DO 5400 I=1,NWMAX
432              if(i.eq.wgprnt) then
433                WRITE(6,320) I
434    320        FORMAT(1H0,2X,'NUMBER OF EMITTING WALL ELEMENT=',I3/)
435                DO 5410 NG=1,NGMAX
436                  GP(NG)=0.0
437   5410        CONTINUE
438                DO 5420 NW=1,NWMAX
439                  IF(I.EQ.NW) THEN
440                    WP(NW)=-1.0
441                  ELSE
442                    WP(NW)=RDWW(I,NW)
443                  END IF
444   5420        CONTINUE
445                CALL PRTDAT
446              end if
447   5400    CONTINUE
448          go to 312
449   5405   continue
450 *---------------------
451 * SELF-ABSORPTION RATIO
452 *---------------------
453          WRITE(6,325)
454    325   FORMAT(1H0,'SELF-ABSORPTION RATIO'/)
455          DO 5422 NG=1,NGMAX
456            GP(NG)=0.0
457   5422   CONTINUE
458          DO 5424 NW=1,NWMAX
459            WP(NW)=ASW(NW)
460   5424   CONTINUE
461          CALL PRTDAT
462        END IF
463 *-------------
464 * TEMPERATURE
465 *-------------
466        WRITE(6,330)
467    330 FORMAT(1H0,'ANALYTICAL RESULTS OF TEMPERATURE (K)'/)
468        DO 5430 NG=1,NGMAX
469          GP(NG)=TG(NG)
470   5430 CONTINUE
471        DO 5440 NW=1,NWMAX
472          WP(NW)=TW(NW)
473   5440 CONTINUE
474        CALL PRTDAT
475 *---------------
476 * WALL HEAT FLUX
```

FIG. 6.11. (*Continued*)

```
477   *----------------
478         WRITE(6,340)
479    340  FORMAT(1H0,'ANALYTICAL RESULTS OF HEAT FLUXES  (W/M**2)'/)
480         DO 5450 NG=1,NGMAX
481            GP(NG)=0.0
482   5450  CONTINUE
483         DO 5460 NW=1,NWMAX
484            WP(NW)=QW(NW)
485   5460  CONTINUE
486         CALL PRTDAT
487         STOP
488         END
489
490
491         SUBROUTINE NXTGAS(NW,NG,IW)
492         COMMON /NTG/INDNXT(4,40),IWMAX,NGMAX
493         NG=1
494         IW=0
495   6000  CONTINUE
496            IW=IW+1
497            IF(IW.GT.IWMAX) THEN
498               IW=1
499               NG=NG+1
500               IF(NG.GT.NGMAX) THEN
501                  WRITE(6,140) NW
502    140           FORMAT(1H ,'WALL ELEMENT (NW=',I3,
503        1           ') IS NOT CONNECTED TO ANY GAS ELEMENTS')
504                  STOP
505               END IF
506            END IF
507         IF(NW.NE.INDNXT(IW,NG)) GO TO 6000
508         RETURN
509         END
510
511
512         SUBROUTINE RANDOM(RAN,RAND)
513         REAL*8 RAND
514         RAND=DMOD(RAND*131075.0D0,2147483649.0D0)
515         RAN=SNGL(RAND/2147483649.0D0)
516         RETURN
517         END
518
519
520         SUBROUTINE PRTDAT
521         COMMON /PRT/GP(40),WP(28)
522         CHARACTER*1 G,W,B15
523         DATA G/'G'/,W/'W'/,B15/'               '/
524   7000  FORMAT(1H ,7(2X,A1,E12.5))
525   7010  FORMAT(1H ,A15,5(2X,A1,E12.5),A15)
526   7020  FORMAT(1H ,7A15)
527   7030  FORMAT(1H ,A15,(2X,A1,E12.5),A15,2(2X,A1,E12.5))
528   7040  FORMAT(1H ,A15,6(2X,A1,E12.5))
529         WRITE(6,7010) B15,(W,WP(I),I=1,5),B15
530         WRITE(6,7000) W,WP(19),(G,GP(J),J=1,5),W,WP(6)
531         WRITE(6,7020) (B15,J=1,7)
532         WRITE(6,7000) W,WP(18),(G,GP(J),J=6,10),W,WP(7)
533         WRITE(6,7030) B15,W,WP(22),B15,(W,WP(J),J=23,24)
534         WRITE(6,7040) B15,W,WP(21),G,GP(12),W,WP(28),W,WP(25),
535        1              G,GP(15),W,WP(8)
536         WRITE(6,7030) B15,W,WP(20),B15,(W,WP(28-J),J=1,2)
537         WRITE(6,7000) W,WP(17),(G,GP(J),J=16,20),W,WP(9)
538         WRITE(6,7020) (B15,J=1,7)
539         WRITE(6,7000) W,WP(16),(G,GP(J),J=21,25),W,WP(10)
540         WRITE(6,7010) B15,(W,WP(16-I),I=1,5),B15
541         RETURN
542         END
```

FIG. 6.11. (*Continued*)

6.3. RADIATIVE HEAT TRANSFER: PROGRAM RADIANW

radiation/absorption being deleted. But lines 338 through 368 are a new addition; they are used to determine the temperature of the isolated solid, B in Fig. 6.12. The value related to convective heat transfer between the gas and the solid CNEWWT (line 347) and all radiative energy incident on the solid (line 350) QRINW are balanced with all radiant heat emitted from the solid, resulting in the determination of the solid temperature. An infinite thermal conductivity of the solid and uniform surface temperature are assumed in the listing example. However, if the actual situation were otherwise, the heat balance equation of each element in the solid would need to be solved to determine the temperature distribution in the solid interior, from which the solid surface temperature distribution is evaluated.

Figure 6.13 presents the solution of the program listing shown in Fig. 6.11. It is seen that (a) the temperature of the solid B is at 855 K, which is higher than the mean temperature of the top and bottom walls; (b) the temperature of the right adiabatic wall diminishes from the bottom to the top; and (c) the upper surface of protrusion A on the left wall has a temperature slightly lower than its neighboring walls because it is not exposed to direct radiation from the bottom wall at 1000 K. Negative values for the wall heat flux of the solid body B signify locally more radiant outflux than influx. The portion of the top wall (at 500 K) that faces enclosed body B is free from direct radiation from the bottom wall but has about the same wall heat fluxes as its neighbors, due to direct radiation

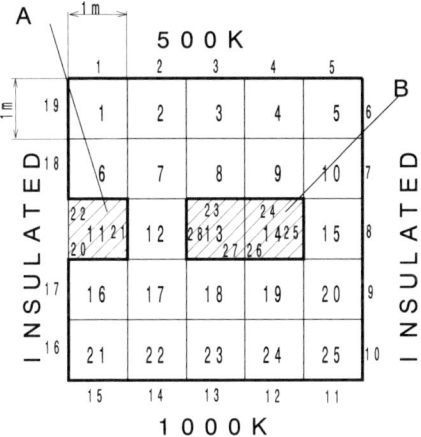

FIG. 6.12. Example of mesh division for RADIANW program

*** COMBINED RADIATION CONVECTION HEAT TRANSFER IN FURNACE ***
       (NON-PARTICIPATING GAS ENVIRONMENT)

NUMBER OF ENERGY PARTICLES EMITTED FROM EACH ELEMENT= 10000

TOTAL INCOMING MASS FLOW= .00000E+00 KG/S/M   AVERAGE HEAT LOAD= .00000E+00 W/M**3

ANALYTICAL RESULTS OF TEMPERATURE (K)

W	.76141E+03	W	.50000E+03	W	.50000E+03	W	.50000E+03	W	.50000E+03	W	.76967E+03
		G	.00000E+00	G	.00000E+00	G	.00000E+00	G	.00000E+00		
W	.75706E+03	G	.00000E+00	G	.00000E+00	G	.00000E+00	G	.00000E+00	W	.79601E+03
		W	.72934E+03	W	.85481E+03	W	.85481E+03	W	.00000E+00		
		W	.85367E+03	W	.85481E+03	W	.85481E+03	W	.00000E+00	W	.85541E+03
		W	.94507E+03	W	.85481E+03	W	.85481E+03	W	.00000E+00		
W	.93431E+03	G	.00000E+00	G	.00000E+00	G	.00000E+00	G	.00000E+00	W	.90313E+03
W	.92393E+03	G	.00000E+00	G	.00000E+00	G	.00000E+00	G	.00000E+00	W	.92287E+03
		W	.10000E+04	W	.10000E+04	W	.10000E+04	W	.10000E+04		

ANALYTICAL RESULTS OF HEAT FLUXES (W/M**2)

W	.00000E+00	W	.94569E+04	W	.10964E+05	W	.11046E+05	W	.10710E+05	W	.11182E+05
		G	.00000E+00	G	.00000E+00	G	.00000E+00	G	.00000E+00		
W	.00000E+00	G	.00000E+00	G	.00000E+00	G	.00000E+00	G	.00000E+00	W	.00000E+00
		W	.00000E+00	W	.00000E+00	W	-.73658E+04	W	-.69800E+04		
		W	.00000E+00	W	.00000E+00	W	-.11902E+03	W	-.26187E+02	W	.00000E+00
		W	.00000E+00	W	.00000E+00	W	.73583E+04	W	.71329E+04		
W	.00000E+00	G	.00000E+00	G	.00000E+00	G	.00000E+00	G	.00000E+00	W	.00000E+00
W	.00000E+00	G	.00000E+00	G	.00000E+00	G	.00000E+00	G	.00000E+00	W	.00000E+00
		W	-.93809E+04	W	-.11074E+05	W	-.10692E+05	W	-.10798E+05	W	-.11415E+05

FIG. 6.13. Output of RADIANW program

from body B. The right end of the top wall has a heat flux higher than its other parts because of reradiation from the right adiabatic wall.

This program is an example of radiative analysis by means of the READ method. A conventional method is available to solve such a problem using the shape factor and radiosity. The latter defines the shape factor F_{ij} to be the fraction of the radiative energy emitted by the surface i that is directly received by the surface j. Then, it determines F_{ij} using equations, diagrams, or the Monte Carlo method. In the evaluation of F_{ij}, effort is directed to only the energy that arrives at the surface j. No consideration is given to whether this energy is partially reflected or is totally absorbed by surface j. When this energy is in the form of reradiation from other surface to the surface j, it is not taken into account. Radiosity W_i is defined as the energy that is emitted from a unit area of the surface i. It can be expressed as

$$W_i = \varepsilon_i \sigma T_i^4 + (1 - \alpha_y)J_i. \tag{6.1}$$

Here, J_i denotes the incident heat flux on the surface i, and α_i is the absorptivity, which is equal to the emissivity ε_i if the surface i is a graybody [Eq. (1.21)]. As shown in Fig. 6.14, the first term on the RHS expresses the emissive power that is proportional to the fourth power of its own temperature, and the second term signifies the reflected component of the incident heat flux.

In the method that uses the shape factor and radiosity, it is necessary to solve simultaneously coupled algebraic equations to determine both the wall temperature and the wall heat flux. Consider heat exchanges between surface i and other $(n - 1)$ surfaces. The net heat loss from the surface i,

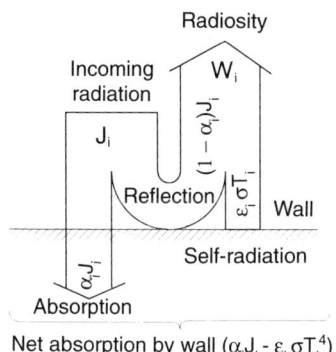

FIG. 6.14. Radiative energy balance at a wall

Q_i, can be expressed as

$$Q_i = \sum_{j=1}^{n} (W_i - W_j) A_i F_{ij}. \tag{6.2}$$

The heat transfer rate of surface i is

$$Q_i = (W_i - J_i) A_i. \tag{6.3}$$

Equations (6.1) and (6.3) are combined to eliminate J_i. It yields

$$Q_i = (\sigma T_i^4 - W_i) A_i \varepsilon_i / (1 - \varepsilon_i). \tag{6.4}$$

The number of Eqs. (6.2) and (6.4) for n surfaces ($i = 1$ to n) is $2n$, in which Q_i, W_i, and T_i of each surface are the variables. With either the wall temperature T_i or the wall heat flux $(Q/A)_i$ of all n surfaces being specified as the boundary conditions, there are $2n$ unknowns in these equations, which can be solved simultaneously. This method is equivalent to the zone method in Section 2.2.2 with the identity of $A_i F_{ij} = \overline{s_i s_j}$, where $\overline{\tau_{ij} s_i s_j}$ is an direct exchange area.

Values obtained with the READ method take some time to evaluate. In the conventional method of using the shape factor and radiosity, the Monte Carlo technique is needed to evaluate the shape factors of a complex system. So there is not much difference in the computational times between the two methods. Because the latter method involves radiosity, it is limited to diffuse surfaces and is not applicable to perfectly reflecting (mirror) surfaces.

6.4. Radiative Heat Transfer in Absorbing–Emitting and Scattering Media

At present, no method other than the Monte Carlo technique can freely treat multidimensional radiative transfer with anisotropic scattering, nonuniformity in properties, and irregularities in system geometry and yet be compatible with finite difference algorithms for solving fluid dynamics. However, the numerical computations of the Monte Carlo method are known to be time-consuming and the Monte Carlo method's results are imperiled by some statistical errors. READ has been developed to overcome these shortcomings. This section demonstrates an application of READ to treat radiative heat transfer in absorbing–emitting and scattering media [28, 29].

6.4. RADIATIVE HEAT TRANSFER IN A–E AND SCATTERING MEDIA

6.4.1. ENCLOSURE

Figure 6.15 shows a two-dimensional rectangular (1- ×1-m) duct. The upper wall is at a higher temperature than the lower wall. Both walls are black. The side walls are adiabatic, gray and diffuse or specular. Both one- and two-dimensional analyses are performed. In the one-dimensional case, both the upper and lower walls are assumed to be isothermal and the side walls are specular. In the two-dimensional case, the upper wall has a stepwise higher uniform temperature region near the center, and the side walls are diffuse.

This rectangular enclosure contains a gray gas with uniformly dispersed gray spherical particles that absorb, emit, and anisotropically scatter the radiative energy [28].

6.4.1.1. *Radiative Characteristics of Wall and Disperse Gas*

Surrounding walls The total radiative energy emitted per unit time from a small gray wall element dA with an emissivity ε_w and a temperature T_w is

$$dQ_{we} = \varepsilon_w \sigma T_w^4 dA, \qquad (6.5)$$

where σ is the Stefan–Boltzmann constant. The corresponding radiative intensity i_{we}, which is independent of direction, is

$$i_{we} = \frac{1}{\pi} \frac{dQ_{we}}{dA}. \qquad (6.6)$$

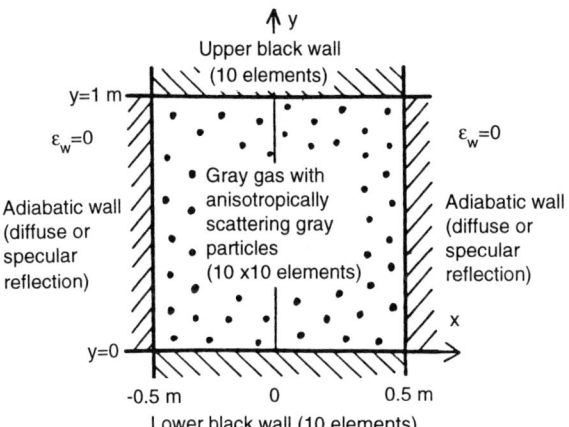

FIG. 6.15. Two-dimensional rectangular duct for radiative transfer analysis

According to Lambert's cosine law, the energy emitted from dA in the direction (θ, η) within a solid angle $d\Omega$, as shown in Fig. 6.16, is

$$d^2Q_{we} = i_{we} \, dA \cos \eta \, d\Omega. \tag{6.7}$$

Out of the radiative energy rate dQ_{wi} incident on dA, only $\varepsilon_w dQ_{wi}$ is absorbed by the wall, while the remaining $dQ_{wr} = (1 - \varepsilon_w) dQ_{wi}$ is reflected. The energy dQ_{wr} reflected in the direction (θ, η) within a solid angle $d\Omega$ can be expressed as

$$d^2Q_{wr} = \frac{1}{\pi} dQ_{wr} \cos \eta \, d\Omega. \tag{6.8}$$

Particles and gas The temperature of the particles is generally assumed to equal that of the surrounding gas T_g. The radiative energy emitted per unit time from a small volume element containing gas and particles is

$$dQ_{ge} = 4a_g \left(dV - \frac{4}{3} \pi R^3 N_s dV \right) \sigma T_g^4 + \varepsilon_s 4\pi R^2 N_s dV \sigma T_g^4$$

$$= 4 \left[a_g \left(1 - \frac{4}{3} \pi R^3 N_s \right) + \varepsilon_s \pi R^2 N_s \right] \sigma T_g^4 dV, \tag{6.9}$$

where a_g denotes the gas absorption coefficient; ε_s, particle emissivity; R, particle radius; and N_s, particle number density. The energy emitted in the direction (θ, η), as shown in Fig. 6.17, within a solid angle $d\Omega$ is given by

$$d^2Q_{ge} = dQ_{ge} \frac{d\Omega}{4\pi}. \tag{6.10}$$

The attenuation of the intensity i of an incident radiation in the s direction, as shown in Fig. 6.18 through a gas volume that contains

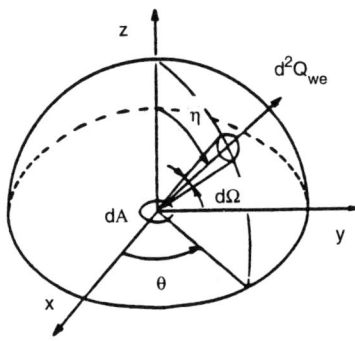

FIG. 6.16. Radiative energy emitted from dA on a solid wall

6.4. RADIATIVE HEAT TRANSFER IN A–E AND SCATTERING MEDIA

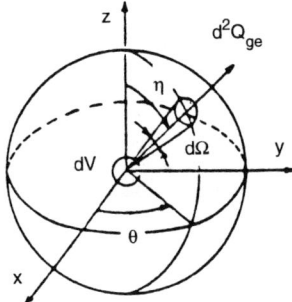

FIG. 6.17. Radiative energy emitted from gas volume dV

dispersed particles due to the absorption of the gas and the adsorption and scattering of the particles, can be expressed by

$$di = -\beta i ds, \quad (6.11)$$

where

$$\beta = a + \sigma_s, \quad (6.12)$$

$$\omega = \sigma_s/\beta, \quad (6.13)$$

and β represents the extinction coefficient; a, total absorption coefficient of gas and particles; σ_s, scattering coefficient; and ω, single scattering albedo. Out of the intensity attenuation di, ωdi is caused by scattering, and the remaining $(1 - \omega)di$ is absorbed. The total absorption coefficient a in Eq. (6.12) can be expressed as

$$a = a_g\left(1 - \tfrac{4}{3}\pi R^3 N_s\right) + \varepsilon_s \pi R^2 N_s \quad (6.14)$$

and includes the radiative equilibrium of emission in Eq. (6.9) and the absorption of dispersed gas volume. The scattering coefficient σ_s reads

$$\sigma_s = (1 - \varepsilon_s)\pi R^2 N_s, \quad (6.15)$$

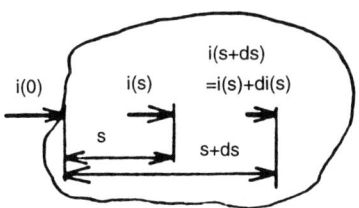

FIG. 6.18. Attenuation of intensity i

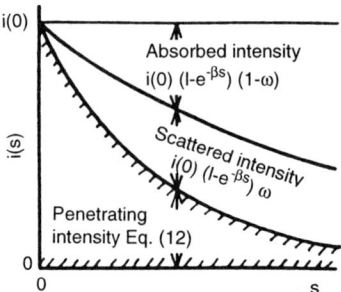

FIG. 6.19. Composition of intensity i

when the radius of the particles is much larger than the wavelength of the radiative energy. Equation (6.11) is integrated as follows:

$$i(s) = i(0)e^{-\beta s}, \qquad (6.16)$$

which is Beer's law. Figure 6.19 depicts the composition of intensity. An anisotropic phase function given by the following equation is used in the present analysis to determine the angle of the scattered energy, as shown in Fig. 6.20:

$$\Phi(\eta) = \frac{8}{3\pi}(\sin \eta - \eta \cos \eta). \qquad (6.17)$$

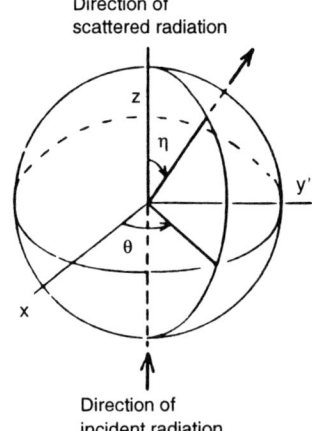

FIG. 6.20. Angle of scattered radiation η

6.4. RADIATIVE HEAT TRANSFER IN A–E AND SCATTERING MEDIA

This expression is graphically illustrated in Fig. 6.21. It is valid for a sphere with diffuse surface and has strong backward scattering characteristics. Equations (6.5) to (6.16) can be used to determine the optical characteristics of dispersed gas as functions of three parameters: β, ω, and Φ or a, σ_s, and Φ.

6.4.1.2. Analysis by Monte Carlo Method

Heat balance The gas volume in Fig. 6.15 is divided into 100 (10 × 10) square elements, while both the upper and lower walls are divided into 10 elements. By prescribing the temperature profiles of the upper and the lower walls as the boundary conditions, the temperature profile in the gas region obtained by the heat balance for a gas element is

$$Q_{\text{out}} = Q_{\text{in}}, \tag{6.18}$$

where Q_{in} and Q_{out} are the net radiative input and output, respectively. The heat adsorbed by each wall element Q_a is

$$Q_a = Q_{\text{out}} - Q_{\text{in}}. \tag{6.19}$$

Note that

$$Q_{\text{out}} = (1 - \alpha) 4 a \sigma T_g^4 \Delta V \tag{6.20}$$

for the gas element and

$$Q_{\text{out}} = (1 - \alpha) \varepsilon_w \sigma T_w^4 \Delta A \tag{6.21}$$

for the wall elements. Here, α is the self-absorption ratio, which represents the ratio of the energy absorbed by the element itself to the total energy emitted from the element. Hence, Q_{out} in both Eqs. (6.20) and

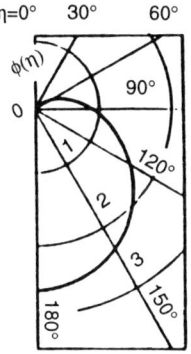

FIG. 6.21. Anisotropic phase function of a gray diffuse sphere

(6.21) represents the total energy emitted from an element and absorbed by other element. The amount of energy emitted from an element I and absorbed by another element J can be expressed by $R_d(I, J) Q_{\text{out}}(I)$. The term $R_d(I, J)$, called READ, is defined as the ratio of the energy emitted from the element I and absorbed by the element J to the total emitted energy $Q_{\text{out}}(I)$. The radiative energy absorbed by an element Q_{in} can be obtained by adding all energy components transferred from all other elements:

$$Q_{\text{in}}(J) = \sum_I R_d(I,J) Q_{\text{out}}(I). \quad (6.22)$$

The wall heat flux reads

$$q = Q_a/\Delta A. \quad (6.23)$$

The radiative properties of the dispersed gas confined within the walls are listed in Table 6.1 cases (a) to (e). Cases (a), (b), and (c) determine the effects of the absorption coefficient under nonscattering conditions. Cases (b), (d), and (e) evaluate the effects of the scattering albedo under a constant extinction coefficient.

To extend a two-dimensional aspect to the rectangular duct system, a 0.2-m-wide higher temperature region at 2500 K is provided at the center of the upper wall at 2000 K. The adiabatic side walls are diffuse surfaces. Both the temperature profile and the wall heat flux are calculated under the same gas conditions as shown in Table 6.1.

Figure 6.22 shows the effects of the absorption coefficient for cases (a) and (c), and Fig. 6.23 depicts the effects of the scattering albedo for cases (b) and (e). These figures reveal that the gas temperature profiles near the higher temperature wall region are very similar, within the ranges of the absorption coefficient and the albedo covered in the computation. Figure 6.24 illustrates the wall heat fluxes for all cases. It is seen that the upper wall heat fluxes fall near the ends of the higher temperature region,

TABLE 6.1. Radiative Properties

Case	Without Scattering			With Anisotropic Scattering		
	(a)	(b)	(c)	(b)	(d)	(e)
$a(\text{m}^{-1})$	1	2	3	2	1.5	1
$\sigma_s(\text{m}^{-1})$	0	0	0	0	0.5	1
$\beta(\text{m}^{-1})$	1	2	3	2	2	2
ω	0	0	0	0	0.25	0.5

6.4. RADIATIVE HEAT TRANSFER IN A−E AND SCATTERING MEDIA 153

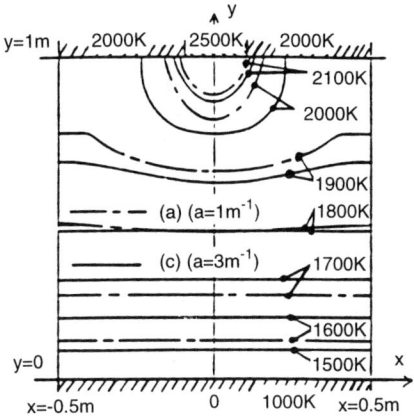

FIG. 6.22. Effects of absorption coefficient on temperature distribution in two-dimensional radiative transfer for nonscattering cases (a) and (c) of Table 6.1

$x = \pm 0.1$ m, which is a typical two-dimensional effect. This is caused by the increase in the gas temperature near the higher temperature wall region in Figs. 6.22 and 6.23. Another two-dimensional effect, as shown in Fig. 6.24, is a rise in the lower wall heat fluxes near the side walls, $x = \pm 0.5$ m. The latter effect is not caused by the higher temperature region of the upper wall, but rather is due to the diffuse side walls. Figure 6.25 compares the temperature distribution in two-dimensional radiative

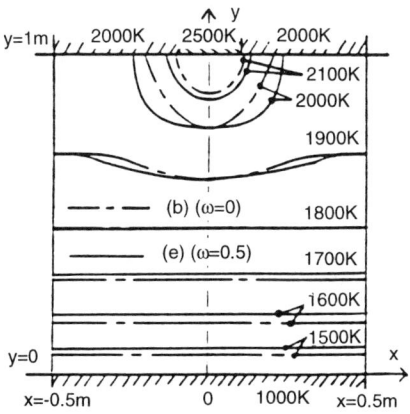

FIG. 6.23. Effects of albedo on temperature distribution in two-dimensional radiative transfer for anisotropic scattering cases (b) and (e) of Table 6.1

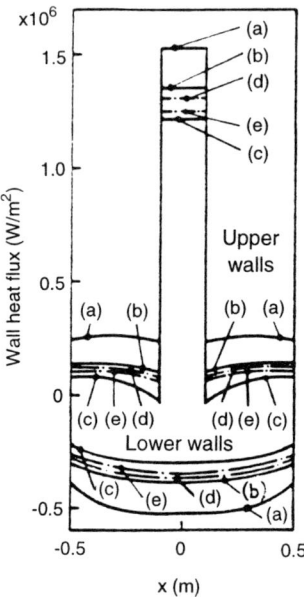

FIG. 6.24. Heat flux distribution in upper and lower walls in two-dimensional radiative transfer for all cases of Table 6.1

heat transfer without scattering, in case (c), and with anisotropic scattering, in case (e). The thermal profiles of both cases nearly coincide. In contrast, Fig. 6.24 shows that the wall heat flux distribution for no scattering case (c), deviates appreciably from that for anisotropic scattering, case (e). It is thus concluded that the radiative heat transfer with anisotropic scattering cannot be simulated by nonscattering radiative transfer with an effective absorption coefficient.

The study concludes the following:

1. An increase in the absorption coefficient and/or the single scattering albedo has adverse effects on radiative heat transfer under anisotropic scattering conditions. It results in an increase in temperature gradient in the gas region and a decrease in wall heat flux.
2. Anisotropic scattering effects cannot be simulated through the use of an effective absorption coefficient.

6.4.2. FURNACE WITH A THROUGHFLOW AND A HEAT-GENERATING (FLAME) REGION

Consider a rectangular duct with a localized heat source, as shown in Fig. 6.26. The upper and lower walls are black and porous and at constant

6.4. RADIATIVE HEAT TRANSFER IN A-E AND SCATTERING MEDIA   155

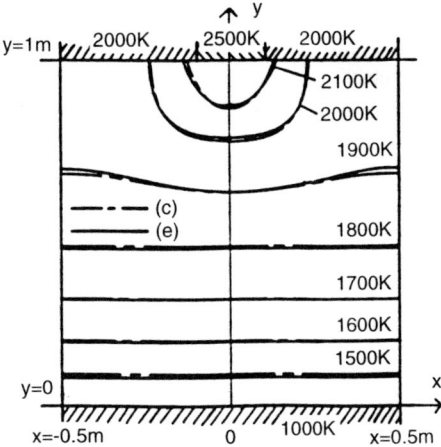

FIG. 6.25. A comparison of temperature distribution in two-dimensional radiative transfer without scattering [case (c) from Table 6.1] and with anisotropic scattering [case (e)]

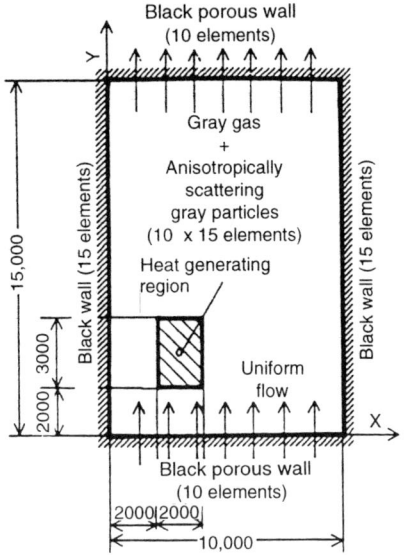

FIG. 6.26. Two-dimensional model for coupled radiation–convection heat transfer analysis

TABLE 6.2. Analytical Conditions

Gas velocity	1.0 m/sec
Heat transfer coeff.	17.4 W/m²-K
Heat generation rate in flame	3.0 MW/m³
Wall emissivity	1.0
Wall temperature	1000 K
Incoming gas temperature	1000 K

TABLE 6.3. Radiative Properties

Case	(a)	(b)	(c)	(d)
$a(m^{-1})$	1.0	0.2	0.4	0.4
$\sigma_s(m^{-1})$	0	0	0.6	0.6
$\beta(m^{-1})$	1.0	0.2	1.0	1.0
ω	0	0	0.8	0.8
Scattering characteristics			Isotropic	Anisotropic

temperature. A uniform gas stream flows vertically through the duct. The operating conditions and radiative properties are list in Tables 6.2 and 6.3, respectively. Note that cases (a) and (b) neglect scattering effects, and cases (c) and (d) take into account the effects of isotropic and anisotropic scattering, respectively. Case (a) has the maximum value for absorption coefficient a.

Figure 6.27 depicts the gas temperature distribution at $X = 3.5$ m in the flow (Y) direction. It indicates that no scattering yields the lowest gas temperatures inside the heat generating region for case (a) and in the non-heat-generating region for case (b). Case (d) with anisotropic scatter-

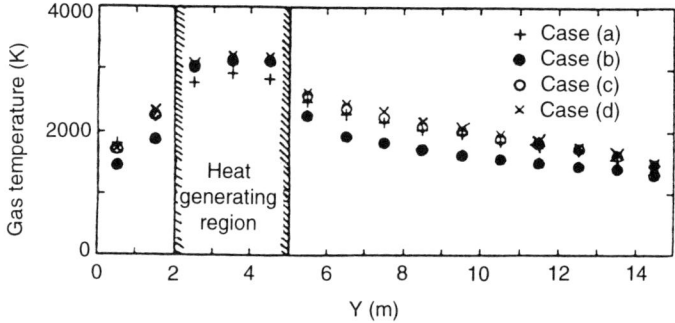

FIG. 6.27. Gas temperature profiles in flow direction ($X = 3.5$ m)

6.4. RADIATIVE HEAT TRANSFER IN A–E AND SCATTERING MEDIA

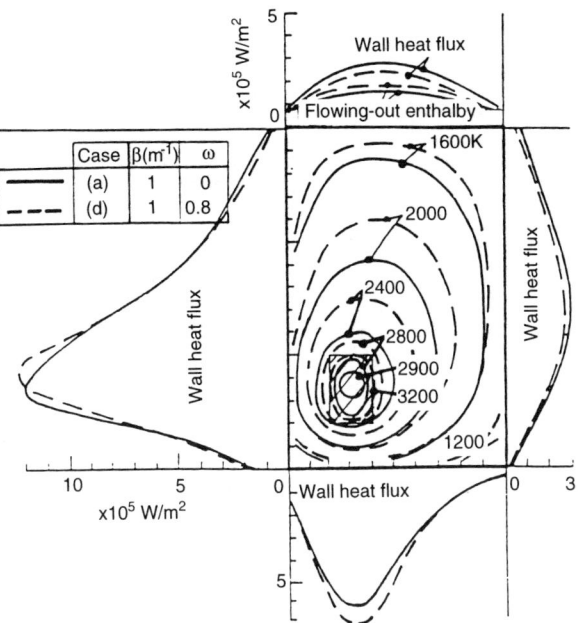

FIG. 6.28. Effects of anisotropic scattering

ing has the highest gas temperature, followed by case (c) with isotropic scattering. Figure 6.28 plots the distributions of wall heat flux, exit gas enthalpy, and gas temperature inside the duct. The peaks of the wall heat flux are seen on both the left side wall and the bottom wall at the locations closest to the heat generating region. In contrast, the maxima of the wall heat flux on the right side wall and the top wall occur at the locations where the temperature of the adjacent gas stream takes a maximum value. The exit gas enthalpy plays a minor role on the total heat exchange in comparison with radiative heat transfer from the gas at a temperature of 2500 to 3000 K. In the case of isotropic scattering, a change in the albedo for single scattering, ω, exerts little effect on the heat transfer. In the case of anisotropic scattering with constant extinction coefficient β, an increase in ω gives rises to the gas temperatures both inside and outside of the heat generating region. It results in a reduction in the wall heat flux and an enhancement of the exit gas enthalpy.

Chapter 7
Some Industrial Applications

7.1. Introduction

Industrial applications of radiative heat transfer are very broad. Some include furnaces in conventional power plants, gas reformers for production of city gases, and combustion chambers in jet engines. During the design and development stages of these heat transfer devices, it is necessary to know the distributions of interior temperatures and wall surface heat fluxes as well as the exit gas temperature. Because of the complexity in the governing equations, boundary conditions, and system geometry, various approximate methods and experiments have been developed and employed. In this chapter, efforts are directed toward the use of the Monte Carlo method in the heat transfer analyses of a simulated boiler, a gas reformer, and the combustion chamber of a jet engine.

7.2. Boiler Furnaces

A concrete example is presented in this section to demonstrate an application of the RADIAN program for analyzing a combined radiative–convective heat transfer problem. Figure 7.1 is a boiler model, which is the object of a heat transfer analysis. It is a two-dimensional (2-D) system of 30 m in height and 6 m in width. A flame is ejected horizontally from burners, which are installed on the left side wall. It is bent upward following the fuel gas flow, forming a flame region enclosed by the hatched lines. The combustion gases exit through the flue outlet, located at the upper right corner of the boiler. The heavy lines enclosing the circumference show the furnace walls. Both the burners and the flue outlet consist of porous walls. The furnace is subdivided into numerous elements for computational purposes, as depicted in Fig. 7.2. Including the dummy gas elements (4, 8, 12) outside the system, there are 40 gas elements and 28 wall elements.

7.2. BOILER FURNACES

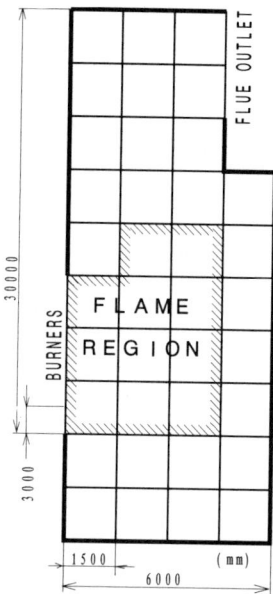

FIG. 7.1. A 2-D boiler model

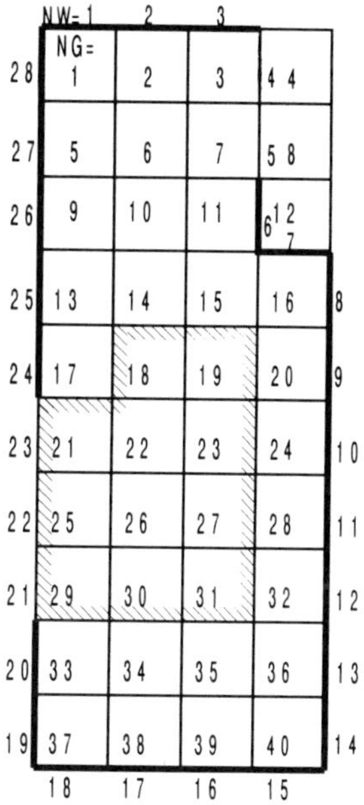

FIG. 7.2. Mesh numbers of gas and wall elements

```
************************************************************
*                                                          *
*                     RADIAN 1                             *
*                                                          *
*      RADIATION- AND CONVECTION-HEAT TRANSFER CALCULATION *
*                  WITHIN AN ENCLOSURE                     *
************************************************************
C
************************************************************
* (SPC): SPECIFICATION STATEMENT
************************************************************
C
      REAL*8 RAND
      COMMON /NTG/INDNXT(4,40),IWMAX,NGMAX
      COMMON /PRT/GP(40),WP(28)
      DIMENSION INDGW(40),INDNT1(4,20),INDNT2(4,20),AK(40),CP(40),
     1          TG(40),QG(40),GMF(4,40),GMF1(4,20),GMF2(4,20),
     2          INDWBC(28),DLW(28),SW(28),TW(28),QW(28),EM(28),H(28),
     3          RDGG(40,40),RDGW(40,28),RDWG(28,40),RDWW(28,28),
     4          ASG(40),ASW(28),S(4),ANEWG(40),BNEWG(40),
     5          ANEWW(28),BNEWW(28)
C
************************************************************
* (IDATA): FIXED AND INITIAL DATA
************************************************************
      DATA NGM,NWM/40,28/
*------------------------
*     DATA FOR GAS ELEMENTS
*------------------------
      DATA INDGW/3*1,0,3*1,0,3*1,0,28*1/
      DATA AK/3*0.2,0.0,3*0.2,0.0,3*0.2,0.0,5*0.2,2*0.8,0.2,3*0.8,
     1        0.2,3*0.8,0.2,3*0.8,9*0.2/
      DATA CP/40*1000./,   CP0/1000./,   TG/40*820./,   TG0/573./
      DATA QG/17*0.,2*4.1E5,0.,13.5E5,6.8E5,4.1E5,0.,13.5E5,6.1E5,
     1        4.1E5,0.,13.5E5,4.1E5,2.7E5,9*0./
      DATA GMF1/
     1    2*0.,-0.1,0.1,      0.1,0.,-4.,3.9,      4.,0.,-6.7,2.7,
     2    4*0.0,              0.,-0.1,-0.2,0.3,    0.2,-3.9,-2.4,6.1,
     3    2.4,-2.7,-6.8,7.1,  4*0.0,               0.,-0.3,-0.1,0.4,
     4    0.1,-6.1,0.2,5.8,   -0.2,-7.1,0.,7.3,    4*0.,
     5    0.,-0.4,-0.3,0.7,   0.3,-5.8,0.,5.5,     0.,-7.3,1.9,5.4,
     6   -1.9,2*0.,1.9,       0.,-0.7,-0.5,1.2,    0.5,-5.5,-0.5,5.5,
     7    0.5,-5.4,0.2,4.7,   -0.2,-1.9,0.,2.1/
      DATA GMF2/
     1    4.5,-1.2,-4.,0.7,   4.,-5.5,-1.6,3.1,    1.6,-4.7,-0.4,3.5,
     2    0.4,-2.1,0.,1.7,    4.5,-0.7,-4.3,0.5,   4.3,-3.1,-2.2,1.0,
     3    2.2,-3.5,-1.2,2.5,  1.2,-1.7,0.,0.5,     4.5,-0.5,-4.1,0.1,
     4    4.1,-1.,-3.1,0.,    3.1,-2.5,-0.6,0.,    0.6,-0.5,0.,-0.1,
     5    0.,-0.1,0.05,0.05,  -0.05,0.,0.04,0.01,  -0.04,0.,0.03,0.01,
     6   -0.03,0.1,0.,-0.07,  0.,-0.05,0.05,0.,    -0.05,-0.01,0.06,0.,
     7   -0.06,-0.01,0.01,0.,-0.07,0.07,2*0./
      DATA INDNT1/
     1 28,1,-2,-5,         -1,2,-3,-6,          -2,3,4,-7,         0,0,0,0,
     2 27,-1,-6,-9,        -5,-2,-7,-10,        -6,-3,5,-11,       0,0,0,0,
     3 26,-5,-10,-13,      -9,-6,-11,-14,       -10,-7,6,-15,      0,0,0,0,
     4 25,-9,-14,-17,      -13,-10,-15,-18,     -14,-11,-16,-19,   -15,7,8,-20,
     5 24,-13,-18,-21,     -17,-14,-19,-22,     -18,-15,-20,-23,   -19,-16,9,-24/
      DATA INDNT2/
     1 23,-17,-22,-25,     -21,-18,-23,-26,     -22,-19,-24,-27,   -23,-20,10,-28,
     2 22,-21,-26,-29,     -25,-22,-27,-30,     -26,-23,-28,-31,   -27,-24,11,-32,
     3 21,-25,-30,-33,     -29,-26,-31,-34,     -30,-27,-32,-35,   -31,-28,12,-36,
     4 20,-29,-34,-37,     -33,-30,-35,-38,     -34,-31,-36,-39,   -35,-32,13,-40,
     5 19,-33,-38,18,      -37,-34,-39,17,      -38,-35,-40,16,    -39,-36,14,15/
      DATA DXG,DYG/1.5,3./
*------------------------
* DATA FOR WALL ELEMENTS
*------------------------
      DATA INDWBC/14*1,4*0,10*1/
      DATA DLW/3*1.5,3*3.,1.5,7*3.,4*1.5,10*3./
      DATA TW/3*670.,2*820.,9*670.,14*670./
      DATA QW/28*0./,  EM/3*0.8,2*1.0,23*0.8/,  H/3*15.,2*0.,23*15./
```

FIG. 7.3. Input data of RADIAN program for a boiler analysis

7.2. BOILER FURNACES

Figures 7.3 and 7.4 present the input data and printout subroutine for the analysis, respectively. They correspond, respectively, to lines 1 through 63 and lines 744 through 760 in the program of Fig. 6.2, which corresponds to the system in Fig. 6.3. Hence, by replacing the portions (lines 1 through 63 and lines 744 through 760) of the program in Fig. 6.2, respectively, with these two parts (Figs. 7.3 and 7.4), the RADIAN program can be used to analyze the boiler system shown in Fig. 7.1.

The following explains, in sequence, the input data in Fig. 7.3. INDGW in line 30 is set equal to 0 for gas elements 4, 8, and 12 because these elements located outside the system are merely dummies. The gas absorption coefficient, AK, in line 31, is 0.8 m^{-1} within the flame region, but is given as 0.2 m^{-1} outside the region. The specific heat of both the gas inside the system and the in-flow gas, CP, CPO = 1000 J/kg K^{-1}. Line 33 gives the initial gas temperature inside the system, TG = 820 K, and the temperature of the incoming gas through the burner section, system TGO = 573 K. In line 34, the heat generation rate within each gas element, QG, is 0 outside the flame. Its value within the flame region increases as the distance to the burners diminishes, as shown in Fig. 7.5. Since the distribution of QG varies with the burner type and operating conditions, its value for the burner selected must be estimated through theory, experiments, or past experience. The mass flow distributions within the GMF1 and GMF2 system, as described in lines 36 through 51, are specified to have a recirculating region at the lower part, as shown in Fig. 7.6. The mass flowing in and flowing out of each gas element must be balanced. This flow distribution can be obtained from experiments on flow models or approximate models for flow analyses. In lines 52 through 63, INDNT1 and INDNT2 are the numbers of those elements that are adjacent to the four sides of each gas element. In line 64, DXG and DYG denote, respectively, the width and the height of a gas element, and are 1.5 m and 3 m, respectively, as seen in Fig. 7.1.

Next, INDWBC in line 68 is an index that identifies the kind of wall-surface boundary conditions: wall temperatures as the boundary conditions of wall elements 1 through 4 and 19 through 28, and wall heat fluxes as the boundary conditions of wall elements 15 through 18. In the present study, DLW in line 69 describes the length of each wall element and it is 1.5 m for the horizontal elements and 3 m in the case of the vertical elements. The wall-surface temperature TW in line 70 is 820 K for elements 4 and 5, which simulate the combustion gas exit, and 670 K for other wall elements. Such a temperature is selected because a superheater is customarily placed at the location of wall elements 4 and 5. The wall-surface heat flux QW in line 71 is 0 W/m^2. Because wall elements 1 through 14 and 19 through 28 are those with the temperature given as the boundary condition, their QW values are merely meaningless dummy ones.

```
750
751
752       SUBROUTINE PRTDAT
753       COMMON /PRT/GP(40),WP(28)
754       CHARACTER*1 G,W,B15
755       DATA G/'G'/,W/'W'/,B15/' '
756  7000 FORMAT(1H ,6(2X,A1,E12.5))
757  7010 FORMAT(1H ,A15,3(2X,A1,E12.5),2A15)
758  7020 FORMAT(1H ,6A15)
759  7030 FORMAT(1H ,4A15,2X,A1,E12.5,A15)
760  7040 FORMAT(1H ,5(2X,A1,E12.5),A15)
761  7050 FORMAT(1H ,A15,4(2X,A1,E12.5),A15)
762       WRITE(6,7010) B15,(W,WP(I),I=1,3),B15,B15
763       WRITE(6,7040) W,WP(28),(G,GP(I),I=1,3),W,WP(4),B15
764       WRITE(6,7020) (B15,I=1,6)
765       WRITE(6,7040) W,WP(27),(G,GP(I),I=5,7),W,WP(5),B15
766       WRITE(6,7020) (B15,I=1,6)
767       WRITE(6,7040) W,WP(26),(G,GP(I),I=9,11),W,WP(6),B15
768       WRITE(6,7030) (B15,I=1,4),W,WP(7),B15
769       DO 8000 I=1,7
770         WRITE(6,7000) W,WP(26-I),(G,GP(8+4*I+J),J=1,4),W,WP(7+I)
771         IF(I.NE.7) THEN
772           WRITE(6,7020) (B15,J=1,6)
773         END IF
774  8000 CONTINUE
775       WRITE(6,7050) B15,(W,WP(19-I),I=1,4),B15
776       RETURN
777       END
```

FIG. 7.4. Printout subroutine of RADIAN program for a boiler analysis

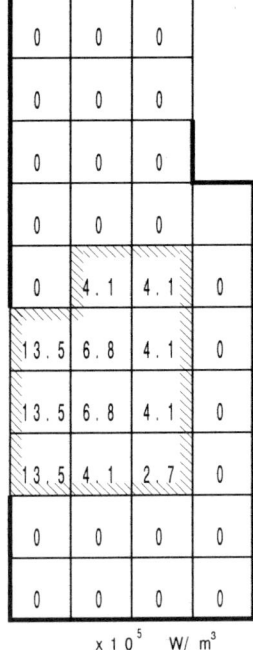

FIG. 7.5. Heat generation rate in gas elements

7.2. BOILER FURNACES

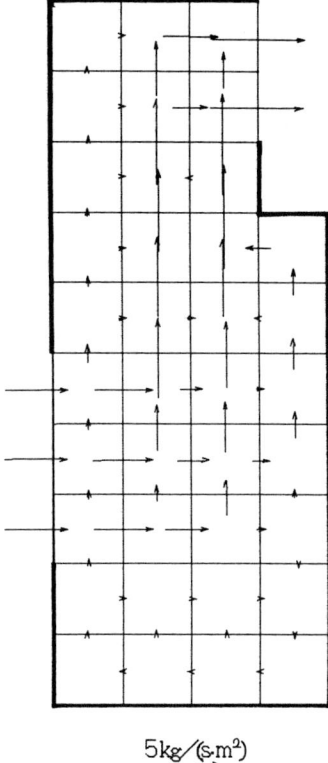

FIG. 7.6. Mass flux distribution

Elements 15 through 18 are those with heat fluxes as the boundary condition by virtue of INDWBC = 1. Hence, by setting QW = 0, these wall elements are defined as the adiabatic walls. The wall-surface emissivity takes a value of unity in the case of elements 4 and 5, which simulate the combustion gas exit. This is because the interiors of tube banks that form a superheater are considered to be cavities. Its value is 0.8 for all other wall elements. The convective heat transfer coefficient between the wall elements and the enclosed gas H is 15 W/m² K^{-1}, except at the combustion gas exit port. In contrast to a finite value of 0.8 for the emissivity of the wall elements at the exit, the local convective heat transfer coefficient of H = 0 suggests no convective heat exchange with the nearby superheater. Note that the gas that flows through the exit cross-section undergoes a radiative heat exchange with the superheater, which is located downstream from the exit cross section. But there is no convective heat exchange between them before the gas flows out through the exit cross section. Hence, the temperature of the gas flowing through the exit cross section is identical to that of the gas entering the superheater.

The solutions obtained from the RADIAN program for the input data of Fig. 7.3 are presented in Figs. 7.7–7.10. They are obtained, not by changing the DATA statements, but by supplying directly the input data through the keyboard as mentioned in Section 6.2. This utilizes the function of the RADIAN program that obtains the approximate solutions for the luminous/nonluminous, or 100%/50% load. Figure 7.7 corresponds to the luminous, 100% load obtained by using those values of the gas absorption,

```
*** COMBINED RADIATION CONVECTION HEAT TRANSFER IN FURNACE ***

NUMBER OF ENERGY PARTICLES EMITTED FROM EACH ELEMENT=  10000
LUMINOUS FLAME (MAXIMUM GAS ABSORPTION COEFF.    =  .800)
FULL LOAD
TOTAL INCOMING MASS FLOW=   .13500E+02 KG/S/M        AVERAGE HEAT LOAD=   .20703E+06 W/M**3
ANALYTICAL RESULTS OF TEMPERATURE (K)
                  W   .67000E+03  W   .67000E+03  W   .67000E+03
W   .67000E+03   G   .11859E+04  G   .14959E+04  G   .14776E+04   W   .82000E+03
W   .67000E+03   G   .13224E+04  G   .16212E+04  G   .16154E+04   W   .82000E+03
W   .67000E+03   G   .14414E+04  G   .17228E+04  G   .16988E+04   W   .67000E+03
                                                                  W   .67000E+03
W   .67000E+03   G   .15682E+04  G   .18388E+04  G   .17915E+04   G   .15809E+04   W   .67000E+03
W   .67000E+03   G   .17202E+04  G   .19912E+04  G   .19693E+04   G   .17006E+04   W   .67000E+03
W   .67000E+03   G   .18115E+04  G   .20613E+04  G   .20005E+04   G   .17460E+04   W   .67000E+03
W   .67000E+03   G   .17896E+04  G   .20057E+04  G   .19694E+04   G   .17503E+04   W   .67000E+03
W   .67000E+03   G   .17277E+04  G   .18608E+04  G   .18306E+04   G   .16030E+04   W   .67000E+03
W   .67000E+03   G   .13898E+04  G   .14649E+04  G   .14554E+04   G   .13711E+04   W   .67000E+03
W   .67000E+03   G   .12490E+04  G   .13106E+04  G   .13140E+04   G   .12422E+04   W   .67000E+03
                  W   .12741E+04  W   .13552E+04  W   .13004E+04   W   .12972E+04
ANALYTICAL RESULTS OF HEAT FLUXES (W/M**2)
                  W   .11006E+06  W   .16378E+06  W   .15568E+06
W   .11612E+06   G   .00000E+00  G   .00000E+00  G   .00000E+00   W   .16108E+06
W   .17907E+06   G   .00000E+00  G   .00000E+00  G   .00000E+00   W   .24984E+06
W   .25398E+06   G   .00000E+00  G   .00000E+00  G   .00000E+00   W   .29604E+06
                                                                  W   .30282E+06
W   .36504E+06   G   .00000E+00  G   .00000E+00  G   .00000E+00   G   .00000E+00   W   .34515E+06
W   .52298E+06   G   .00000E+00  G   .00000E+00  G   .00000E+00   G   .00000E+00   W   .53290E+06
W   .55586E+06   G   .00000E+00  G   .00000E+00  G   .00000E+00   G   .00000E+00   W   .57495E+06
W   .49208E+06   G   .00000E+00  G   .00000E+00  G   .00000E+00   G   .00000E+00   W   .56473E+06
W   .40887E+06   G   .00000E+00  G   .00000E+00  G   .00000E+00   G   .00000E+00   W   .40553E+06
W   .21578E+06   G   .00000E+00  G   .00000E+00  G   .00000E+00   G   .00000E+00   W   .21874E+06
W   .13796E+06   G   .00000E+00  G   .00000E+00  G   .00000E+00   G   .00000E+00   W   .14399E+06
                  W   .00000E+00  W   .00000E+00  W   .00000E+00   W   .00000E+00
```

FIG. 7.7. Analytical results for the heat transfer in a boiler (luminous flame, full load)

7.2. BOILER FURNACES

heat generation rate, and flow velocity distributions within the flame region, as given in the DATA statements of Fig. 7.3. Figure 7.8 corresponds to the nonluminous, 100% load, with the absorption coefficient both inside and outside the flame being 0.2 m^{-1}. Figure 7.9 shows the results for the luminous, 50% load case, with the same absorption coefficient distribution as Fig. 7.7, but both the heat generation and flow velocity distributions are reduced by one-half. In all three figures, the upper part presents the temperature distribution, and the lower one shows the wall-surface heat flux distribution.

Figure 7.10 presents the temperature distributions in the contour line expression for the three cases of Figs. 7.7–7.9. A comparison of Figs. 7.7 and 7.8 with Figs. 7.10(a) and (b) disclosed that when the flame becomes

```
*** COMBINED RADIATION CONVECTION HEAT TRANSFER IN FURNACE ***

NUMBER OF ENERGY PARTICLES EMITTED FROM EACH ELEMENT=  10000
NON-LUMINOUS FLAME (MAXIMUM GAS ABSORPTION      COEFF. =   .200)
FULL LOAD
  TOTAL INCOMING MASS FLOW=  .13500E+02 KG/S/M      AVERAGE HEAT LOAD= .20703E+06 W/M**3
ANALYTICAL RESULTS OF TEMPERATURE (K)
              W   .67000E+03   W   .67000E+03   W   .67000E+03
   W .67000E+03  G   .12084E+04   G   .15246E+04   G   .15073E+04   W   .82000E+03
   W .67000E+03  G   .13563E+04   G   .16603E+04   G   .16593E+04   W   .82000E+03
   W .67000E+03  G   .14614E+04   G   .17712E+04   G   .17563E+04   W   .67000E+03
                                                                    W   .67000E+03
   W .67000E+03  G   .15660E+04   G   .19017E+04   G   .18623E+04   G   .15797E+04   W   .67000E+03
   W .67000E+03  G   .16885E+04   G   .21009E+04   G   .21108E+04   G   .16754E+04   W   .67000E+03
   W .67000E+03  G   .18136E+04   G   .21185E+04   G   .21183E+04   G   .17226E+04   W   .67000E+03
   W .67000E+03  G   .18007E+04   G   .20676E+04   G   .20724E+04   G   .17432E+04   W   .67000E+03
   W .67000E+03  G   .17664E+04   G   .19402E+04   G   .19452E+04   G   .16246E+04   W   .67000E+03
   W .67000E+03  G   .13850E+04   G   .14650E+04   G   .14543E+04   G   .13836E+04   W   .67000E+03
   W .67000E+03  G   .12513E+04   G   .13219E+04   G   .13146E+04   G   .12478E+04   W   .67000E+03
                 W   .12940E+04   W   .13298E+04   W   .13283E+04   W   .12911E+04
ANALYTICAL RESULTS OF HEAT FLUXES (W/M**2)
              W   .12328E+06   W   .17648E+06   W   .16678E+06
   W .12624E+06  G   .00000E+00   G   .00000E+00   G   .00000E+00   W   .17592E+06
   W .19711E+06  G   .00000E+00   G   .00000E+00   G   .00000E+00   W   .27795E+06
   W .27023E+06  G   .00000E+00   G   .00000E+00   G   .00000E+00   W   .32495E+06
                                                                    W   .32302E+06
   W .35492E+06  G   .00000E+00   G   .00000E+00   G   .00000E+00   G   .00000E+00   W   .35232E+06
   W .45821E+06  G   .00000E+00   G   .00000E+00   G   .00000E+00   G   .00000E+00   W   .47636E+06
   W .54136E+06  G   .00000E+00   G   .00000E+00   G   .00000E+00   G   .00000E+00   W   .50498E+06
   W .50137E+06  G   .00000E+00   G   .00000E+00   G   .00000E+00   G   .00000E+00   W   .48709E+06
   W .41666E+06  G   .00000E+00   G   .00000E+00   G   .00000E+00   G   .00000E+00   W   .36827E+06
   W .21720E+06  G   .00000E+00   G   .00000E+00   G   .00000E+00   G   .00000E+00   W   .21544E+06
   W .14148E+06  G   .00000E+00   G   .00000E+00   G   .00000E+00   G   .00000E+00   W   .13811E+06
                 W   .00000E+00   W   .00000E+00   W   .00000E+00   W   .00000E+00
```

FIG. 7.8. Analytical results for the heat transfer in a boiler (nonluminous flame, full load)

```
*** COMBINED RADIATION CONVECTION HEAT TRANSFER IN FURNACE ***

NUMBER OF ENERGY PARTICLES EMITTED FROM EACH ELEMENT=  10000
LUMINOUS FLAME (MAXIMUM GAS ABSORPTION COEFF.    =  .800)
HALF LOAD
TOTAL INCOMING MASS FLOW=  .67500E+01 KG/S/M        AVERAGE HEAT LOAD=  .10351E+06 W/M**3
ANALYTICAL RESULTS OF TEMPERATURE (K)
              W  .67000E+03  W  .67000E+03  W  .67000E+03
W .67000E+03  G  .10091E+04  G  .12515E+04  G  .12356E+04  W  .82000E+03
W .67000E+03  G  .11225E+04  G  .13647E+04  G  .13586E+04  W  .82000E+03
W .67000E+03  G  .12293E+04  G  .14594E+04  G  .14355E+04  W  .67000E+03
                                                          W  .67000E+03
W .67000E+03  G  .13485E+04  G  .15720E+04  G  .15278E+04  G  .13458E+04  W  .67000E+03
W .67000E+03  G  .15110E+04  G  .17247E+04  G  .16948E+04  G  .14627E+04  W  .67000E+03
W .67000E+03  G  .16523E+04  G  .18189E+04  G  .17376E+04  G  .15103E+04  W  .67000E+03
W .67000E+03  G  .16412E+04  G  .17799E+04  G  .17205E+04  G  .15176E+04  W  .67000E+03
W .67000E+03  G  .15827E+04  G  .16534E+04  G  .16000E+04  G  .13921E+04  W  .67000E+03
W .67000E+03  G  .12459E+04  G  .13062E+04  G  .12903E+04  G  .12048E+04  W  .67000E+03
W .67000E+03  G  .11146E+04  G  .11701E+04  G  .11714E+04  G  .11013E+04  W  .67000E+03
              W  .11393E+04  W  .12094E+04  W  .11586E+04  W  .11534E+04
ANALYTICAL RESULTS OF HEAT FLUXES (W/M**2)
              W  .54955E+05  W  .82396E+05  W  .79092E+05
W .58179E+05  G  .00000E+00  G  .00000E+00  G  .00000E+00  W  .69473E+05
W .91297E+05  G  .00000E+00  G  .00000E+00  G  .00000E+00  W  .11548E+06
W .13303E+06  G  .00000E+00  G  .00000E+00  G  .00000E+00  W  .15433E+06
                                                          W  .16325E+06
W .19927E+06  G  .00000E+00  G  .00000E+00  G  .00000E+00  G  .00000E+00  W  .18569E+06
W .30461E+06  G  .00000E+00  G  .00000E+00  G  .00000E+00  G  .00000E+00  W  .29441E+06
W .36818E+06  G  .00000E+00  G  .00000E+00  G  .00000E+00  G  .00000E+00  W  .32750E+06
W .33774E+06  G  .00000E+00  G  .00000E+00  G  .00000E+00  G  .00000E+00  W  .32758E+06
W .28012E+06  G  .00000E+00  G  .00000E+00  G  .00000E+00  G  .00000E+00  W  .23726E+06
W .13662E+06  G  .00000E+00  G  .00000E+00  G  .00000E+00  G  .00000E+00  W  .13113E+06
W .85125E+05  G  .00000E+00  G  .00000E+00  G  .00000E+00  G  .00000E+00  W  .87728E+05
              W  .00000E+00  W  .00000E+00  W  .00000E+00  W  .00000E+00
```

FIG. 7.9. Analytical results for the transfer in a boiler (luminous flame, half load)

nonluminous, the flame temperature increases due to a reduction in the heat dissipation by radiation from the flame. It is accompanied by an increase in the gas temperature at the furnace exit. The nonluminous case has a lower heat flux in the region near the flame that is subject to direct radiation from the flame. The wall heat flux, at a distance from the flame, is higher in the case of a nonluminous flame due to an elevation in the gas temperature. The 50% load case, with both the heat generation rate and flow velocity inside the flame being reduced by one-half, should, in principle, have an identical temperature distribution as the 100% load case, in the absence of heat transfer to the wall surface. In reality, however, the former is lower in both the gas temperature and the heat flux than the latter, due to radiation and convection to the wall surface. This is

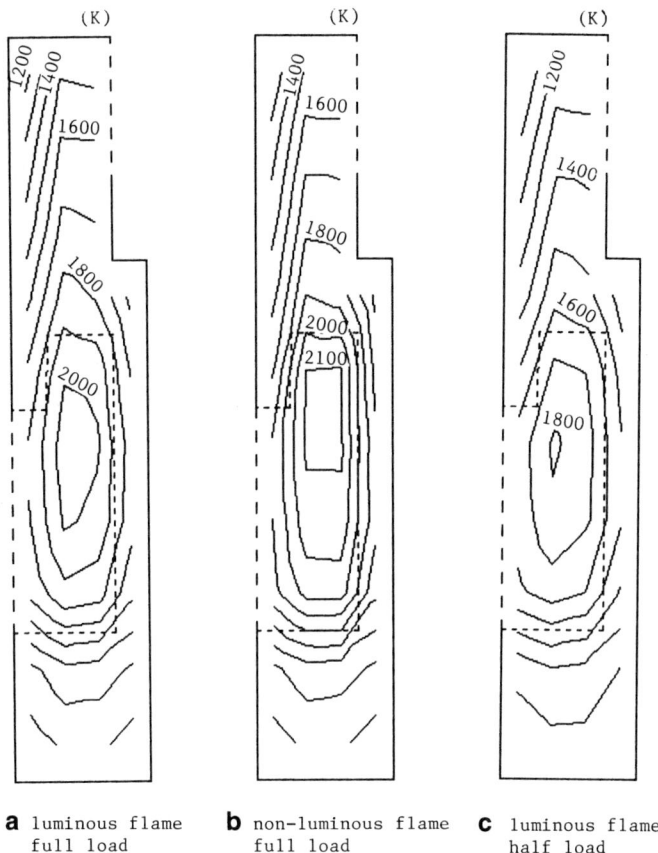

a luminous flame full load **b** non-luminous flame full load **c** luminous flame half load

FIG. 7.10. Temperature profiles in a boiler model: (a) luminous flame full load, (b) nonluminous flame full load, and (c) luminous flame half load

revealed through a comparison among Figs. 7.7 and 7.9 with Figs. 7.10(a) and (c).

7.3. Gas Reformer

A furnace that produces gases rich in hydrogen through the following dissociation processes is called a gas reformer [30]:

$$C_nH_n + mH_2O \rightarrow mCO + (m + n/2)H_2, \qquad (7.1)$$

$$CO + H_2O \rightleftarrows CO_2 + H_2. \qquad (7.2)$$

Because the reaction is an endothermic one, it is necessary to heat the catalytic part through which the row material gas flows. Figure 7.11 is a schematic of a cylindrical-type gas reformer. It consists of reactor tubes and a furnace part. The raw gas is reformed by the catalyst in the reactor tubes. To maintain the reaction region at a desired temperature level, the outer surface of the reactor tubes is heated by a burner flame. The reforming performance is strongly influenced by the temperature of the reactor tubes within which chemical reactions by the catalyst take place. The heat transfer to the reactor tubes is dominated by radiation because the surrounding temperature of the reaction region is high, about 1000 K. The Monte Carlo method was employed to investigate the three-dimensional (3-D) combined radiative–convective heat transfer in order to determine the characteristics of the gas reformer. This analysis, in combination with a simulation of the chemical reactions within reactor tubes, leads to the determination of temperature distribution on the reactor tube surface, which serves as the boundary condition for the heat transfer analysis within the furnace.

Figure 7.12 shows a model for gas reformer analysis. Due to the axisymmetry of the furnace, only one-half the pitch of reactor tubes in the circumferential direction needs to be analyzed. The radiation on the

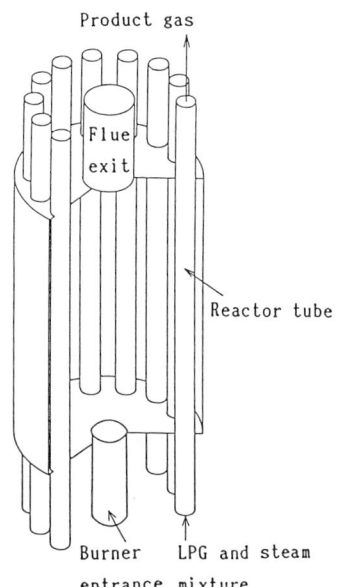

FIG. 7.11. Cylindrical gas reformer

7.3. GAS REFORMER

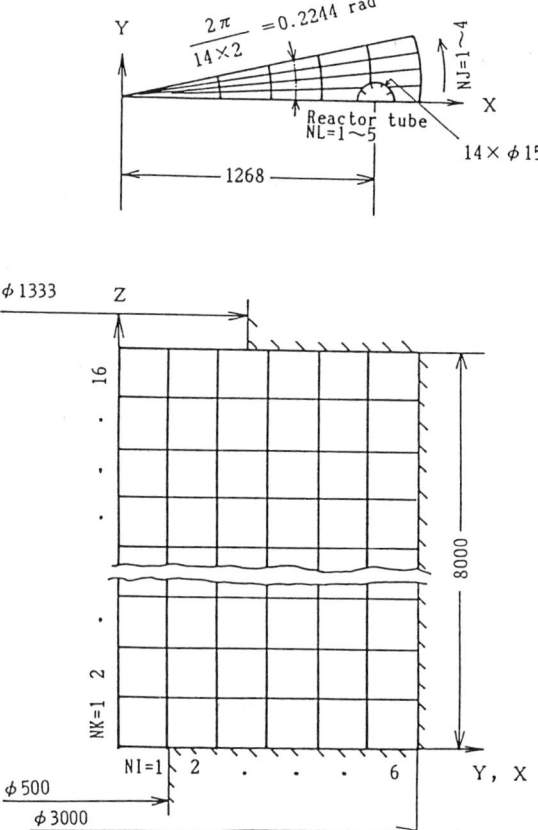

FIG. 7.12. Mesh division of cylindrical gas reformer furnace

divided surface in the circumferential direction is perfectly reflected due to symmetry.

Figure 7.13 presents three flow models at the burner exit with the same inlet flow velocity of $V = 14$ m/sec. Models (b) and (c) have three burner injection angles each: $\theta_1/2$, $\theta_2/2$, and $\theta_3/3$. Figure 7.14 shows the mass flux distributions inside gas reformers, which are obtained using the finite volume method with the standard k-ε model. The results are then utilized to calculate the enthalpy transport terms in the energy equation in Chapter 3.

Regarding the combustion reactions within the furnace, the flame shapes, shown as the dotted regions in Fig. 7.15, are obtained from past experi-

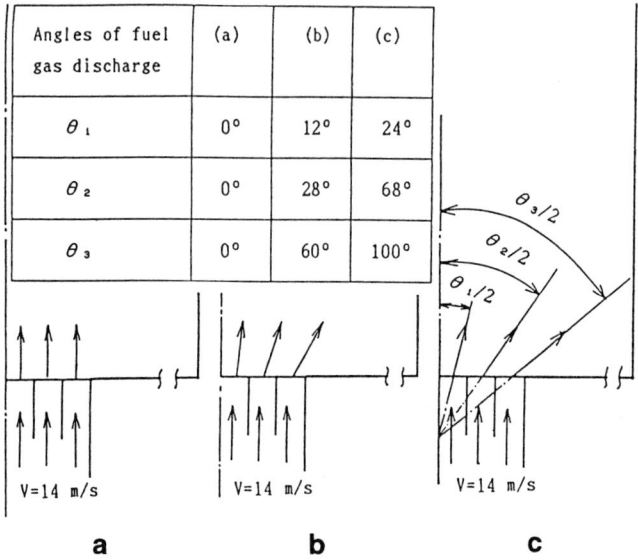

FIG. 7.13. Burner models

ence. Each figure has its corresponding flow fields in Fig. 7.14, each of which, in turn, has its corresponding burner injection angle. The following equation by Wiebe describes the heat generation rate distribution along streamlines:

$$Q_h = 1 - \exp\{-6.9[t_1(t/t_z)^{m+1}]\}. \qquad (7.3)$$

Here, Q_h denotes the volumetric rate of heat generation; t_z, time required for the fluid to travel from the burner exit along a streamline to the outer boundary of the flame; and t_1 and m, coefficients. The values of $t_1 = 0.8$ and $m = -0.1$ are selected in order to match the distribution of Q_h with the results of other combustion analyses. Equation (7.3) is used to simulate the heat generation by combustion inside the furnace.

The thermal boundary conditions and furnace dimensions used in the analysis are listed in Table 7.1. The analysis begins with a postulation of the wall temperature distribution in the axial direction of the reactor tubes. The Monte Carlo method is then utilized to determine the temperature distribution of combustion gases within the furnace and the heat flux distribution in the axial direction of the reactor tubes. It is then followed by solving the chemical reactions within the reactor tubes using the wall heat flux distribution of the reactor tubes as the boundary conditions.

7.3. GAS REFORMER

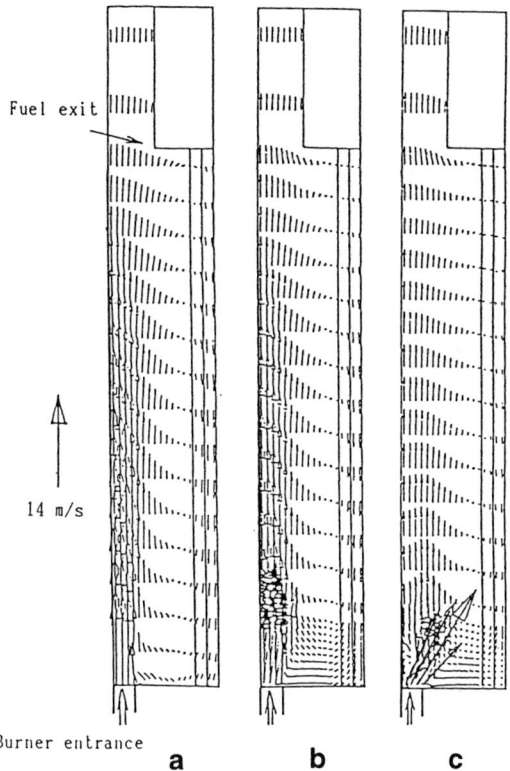

FIG. 7.14. Velocity profiles in the furnace for different burner models

Subsequently, the wall temperature distribution of reactor tubes is evaluated. This procedure is repeated for the heat transfer analysis of the cylindrical-type gas reformer.

Figures 7.15 and 7.16 present the temperature distribution in gas reformers and the heat flux distribution in the axial direction of reactor tubes, respectively. They include the results for the three burner injection angles.

Figure 7.15(a) shows that when the burner injection angle is zero or small, the high-temperature region downstream from the location of the maximum temperature on the central axis is elongated in the downstream direction due to the rapid flow stream. The temperatures in the vicinity of the right-side reactor tubes at a distance from the central axis are much lower than those on the central axis. In comparison, when the burner injection angle is large, in Fig. 7.15(c), those temperatures on the central

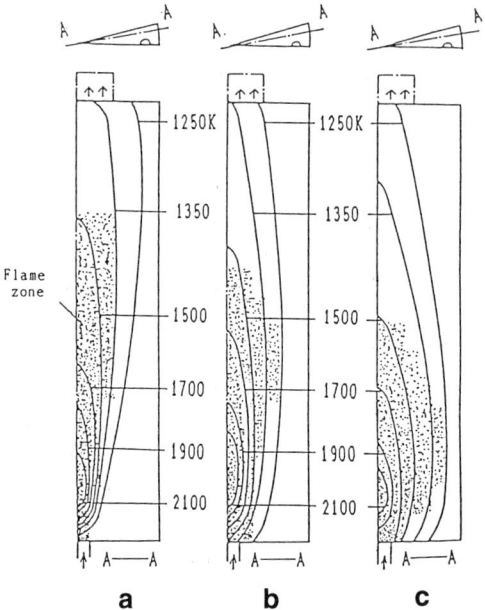

FIG. 7.15. Temperature distributions in furnace for different burner models

TABLE 7.1.
THERMAL BOUNDARY CONDITIONS AND FURNACE DIMENSIONS

Furnace radius	1500	mm
Furnace height	8000	mm
Number of tubes	14	
Distance between tubes and furnace center	1268	mm
Number of elements		
Radial	6	
Circumference	4	
Axial	16	
Tube surface	5	
Absorption coefficient		
Combustion gas	0.22	1/m
Flame	0.80	1/m
Heat transfer coefficient by convection	20	$W/m^2\ K^{-1}$
Special heat of gas	1090	$J/kg\ K^{-1}$
Mass velocity of gas	0.1	$kg/m^2\ sec^{-1}$
Air inlet temperature	550	K
Emissivity		
Furnace wall	0.3	
Reactor tube wall	0.8	
Net heat flux of furnace wall	0 (adiabatic)	kW
Heat generation in flame	1.715×10^2	kW/m^3

7.4. COMBUSTION CHAMBERS IN JET ENGINES

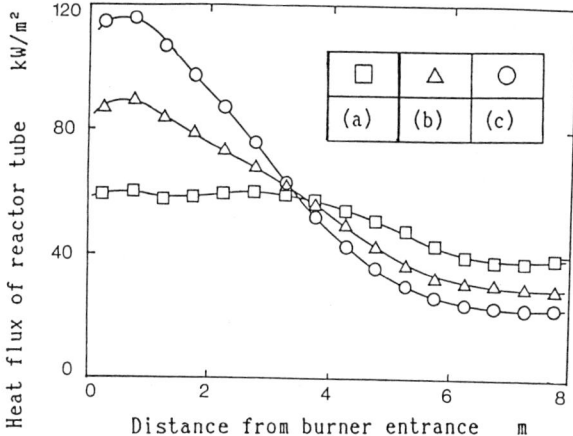

FIG. 7.16. Effect of burner types on heat flux

axis near the burners are almost the same as those in the case of small burner injection angles. But the high-temperature region is expanded in the horizontal direction, resulting in a higher average temperature on the horizontal cross section. Hence, the tube heat flux in the vicinity of the burners becomes high, as depicted in Fig. 7.16 for the large burner injection angle case, case (c). The reason is that the larger the burner injection angle, the broader the flame of the high absorption coefficient, It results in increases in both radiative effects and tube heat flux. As the burner injection angle is increased, as seen in Fig. 7.15(c), the gas temperature surrounding the heat transfer tubes in the vicinity of the burners rises. The convective heat transfer also increases, but accounts for only about 1/20 of the radiative component, thus contributing little to an enhancement in the heat flux.

Figures 7.17 depicts the analytical results of both the temperature and the heat flux distributions in the axial direction of the reactor tubes, corresponding to the burner injection angle of Fig. 7.13(b). We see that the high heat flux at the entrance to the reactor tubes (left side in Fig. 7.17) compensates for a fall in the temperature in the region of strong endothermic reactions. Hence, the reactor tube surface temperature is relatively uniform in the axial direction of reactor tubes.

7.4. Combustion Chambers in Jet Engines

To enhance thermal efficiency, the inlet gas temperature to gas turbines in aircraft engines has been increased. In keeping with this tendency, it is

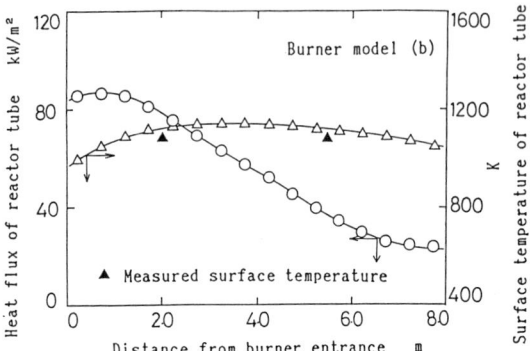

FIG. 7.17. Comparison of heat transfer simulation and experimental results

desirable to improve the accuracy in the evaluation of the radiative heat transfer from the combustion chamber to the first-stage nozzle vanes in the high-pressure turbine [9]. The Monte Carlo method has been employed to investigate the radiative heat flux distributions in the nozzle vane and rotor blades, using both 2-D and 3-D models. Also being evaluated are the effects of various parameters on the heat flux distributions.

Figures 7.18 and 7.19 depict the 2-D and 3-D models of the combustion chamber, respectively. Both models 1 and 2 are 2-D, but have different nozzle bend shapes: One is straight, the other curved. The element subdivisions of models 1, 2, and 3 are depicted in Fig. 7.20. The analytical models are illustrated in Fig. 7.21 The primary air enters the combustion chamber through a porous flat plate (entrance wall), which simulates the burners at a mass flux of G_1 per unit thickness perpendicular to the paper and a temperature of T_1. Heat generation occurs in the flame zone at a heat load of Q_1. The air temperature falls in the dilution zone where it mixes with the secondary air at a mass flux of G_2 and a temperature of T_2. The resulting stream flows into the passage between the parallel, plate-type nozzle vanes inclined at an angle ξ. To simulate the static temperature drop in the gas accelerating through the passage, an equivalent heat absorption term Q_3 is distributed in the gas. The situation is treated as a case of turbulent flow over a flat plate such that the heat transfer coefficient on the nozzle vane surfaces may diminish in proportion to the -0.2 power of the distance along the flow, and such that its average values over the entire length of the nozzle vanes may agree with the conventional, experimental data. Since the heat transfer coefficient distribution along the nozzle vanes is specified, the mass velocity distribution is to be used only in the enthalpy transport terms of the energy equations [Eqs. (3.7) and (3.8)] in Section 3.2. Hence, a uniform flow pattern parallel to the side

7.4. COMBUSTION CHAMBERS IN JET ENGINES

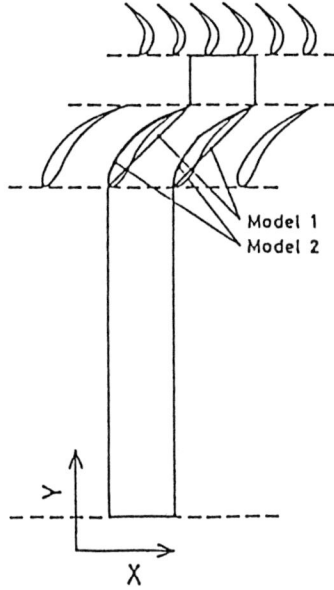

FIG. 7.18. Top view of models 1 and 2

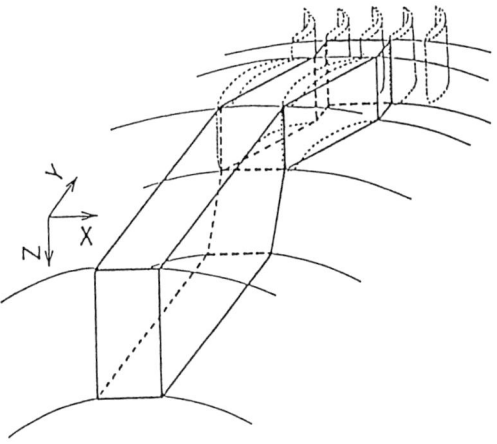

FIG. 7.19. Overview of model 3

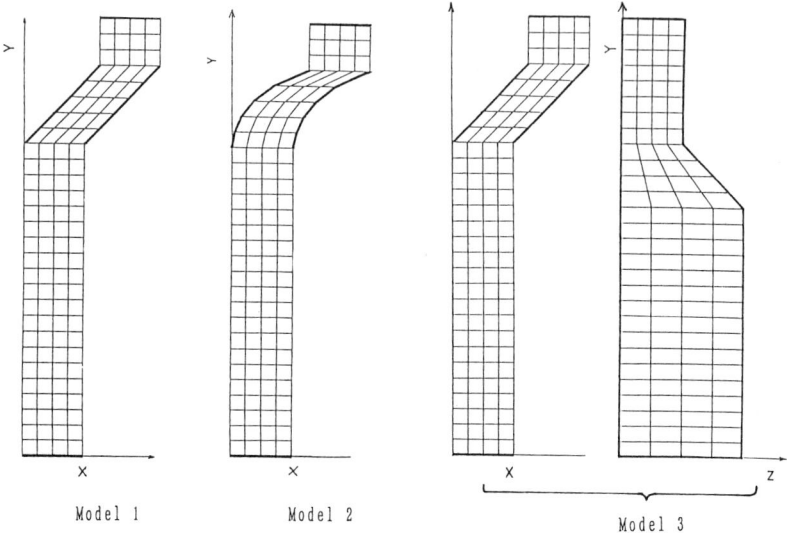

Fig. 7.20. Mesh divisions

walls is assumed in the flow field. It is considered from the viewpoint of radiative heat transfer that the model in Fig. 7.21 is connected on both sides repeatedly to an infinite number of identical models. The solid line represents a gray wall and the gas is a gray body. Because of periodicity on both sides, the radiation leaving the model through the boundary indicated

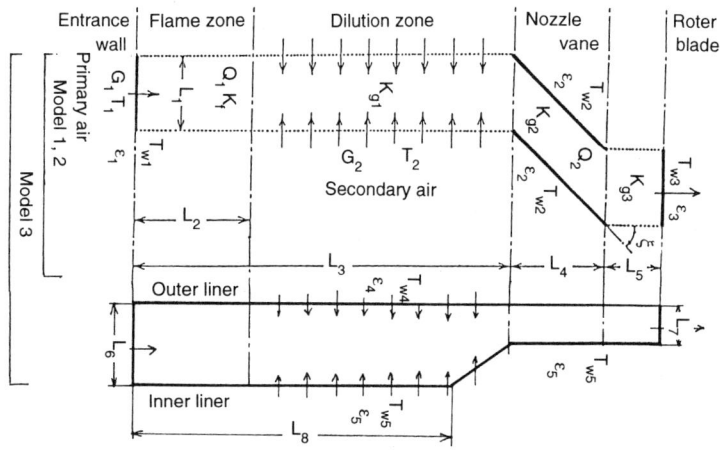

Fig. 7.21. Analtyical models

7.4. COMBUSTION CHAMBERS IN JET ENGINES

by broken lines has its counterpart, which enters the model from the opposite side through the boundary with the same angle. In other words, the periodic boundary condition is adopted.

Table 7.2 lists the operating conditions and system dimensions, which are selected in reference to the actual cases. The values inside parentheses define the range of variation of the corresponding parameter. The values outside parentheses are used as the basic conditions to evaluate the effects of that parameter. Using a HITAC-M682H computer, the computational time required for each element emitting 50,000 energy particles is 16 minutes for both models 1 and 2, and 100 minutes for model 3.

Figure 7.22 shows the temperature distribution in the combustion chamber of model 1 under the basic conditions. The figure shows that the maximum temperature in the flame region is 2200 K; gas temperature at the nozzle vane inlet, 1700 K; and rotor blade inlet temperature, 1400 K. The results are close to the operating conditions in actual engines. The corresponding heat flux distribution on the nozzle vane surface is presented in Fig. 7.23. It is found that radiative heat transfer is about 10% of

TABLE 7.2.
OPERATING CONDITIONS AND SYSTEM DIMENSIONS

Primary air flame	Solid walls
G_1, 33.0 kg/m sec^{-1}	ε_1, 1.0
	ε_2, 0.8 (0.6–1.0)
T_1, 670 K	ε_3, 1.0
	ε_4, 0.8
	ε_5 0.8
p_1, 1.52 MPa	T_{w1}, 1100 K
	T_{w2} 1270 K
Q_1, 890 MW/m^3	T_{w3}, 1100 K
K_f, 6.0 1/m (1–40)	T_{w4}, T_{w5}, 1270 K (800–1270)
Secondary air combustion gas	Dimensions
G_2, 20.0, kg/m sec^{-1} (11–34)	L_1, 40 mm.
	L_2, 60 mm
T_2, 670 K	L_3, 200 mm
	L_4, 50 mm
p_2, 0.91 MPa	L_5, 30 mm
	L_6, 16 mm
Q_2, 220 MW/m^3 (model 1.2); 440 MW/m^3 (model 3)	L_7, 8mm
	L_8, 160 mm
K_{a1} 1.0 1/m	ξ, 45°(0°–60°)
K_a 0.6 1/m	

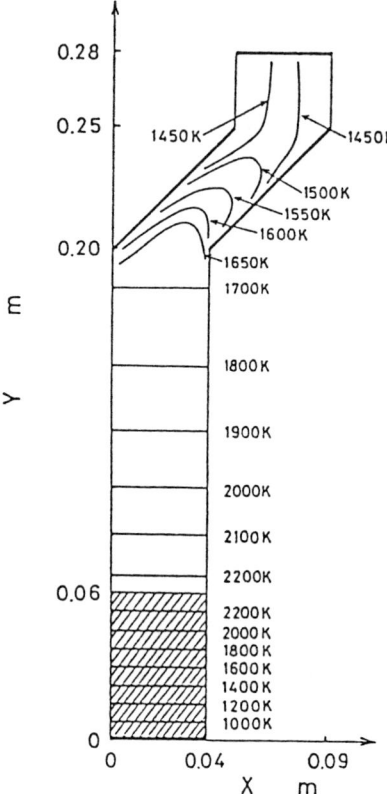

FIG. 7.22. Temperature profile in the combustion chamber (model 1)

the convective heat transfer at the front end and that the heat flux on the pressure side is about three times that on the suction side. The radiative heat flux takes a negative value at the rear end of the suction side, which faces the rotor blades at a lower temperature. Figure 7.24 depicts the radiative heat flux distributions on the nozzle vane surfaces of models 1, 2, and 3 in Fig. 7.20. The temperatures of the liners sandwiching the nozzle vane from top and bottom, T_{w4} and T_{w5} (see Fig. 7.21), are lower than that of the gas flowing through the nozzle vane. Model 3 includes the radiative cooling effect of the liners, but models 1 and 2 do not. Hence, the radiative heat flux of model 3 is lower than that of models 1 and 2.

Figure 7.25 shows the distribution of radiative heat fluxes to the combustion chamber liners, which constitute the left and right boundaries of the domain on the right side of Fig. 7.21. The effects of angle ξ on the

7.4. COMBUSTION CHAMBERS IN JET ENGINES

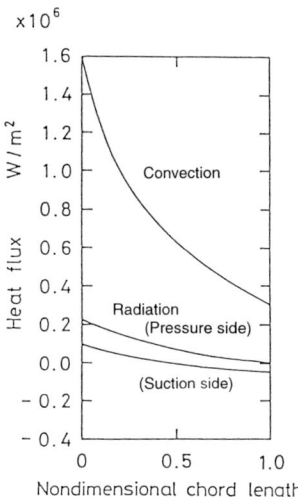

FIG. 7.23. Heat flux along nozzle vane surface

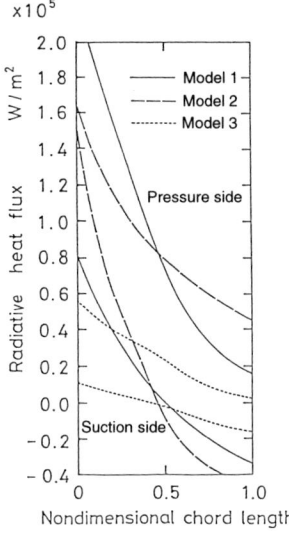

FIG. 7.24. Three-dimensional effects on radiative heat flux along nozzle vane surface

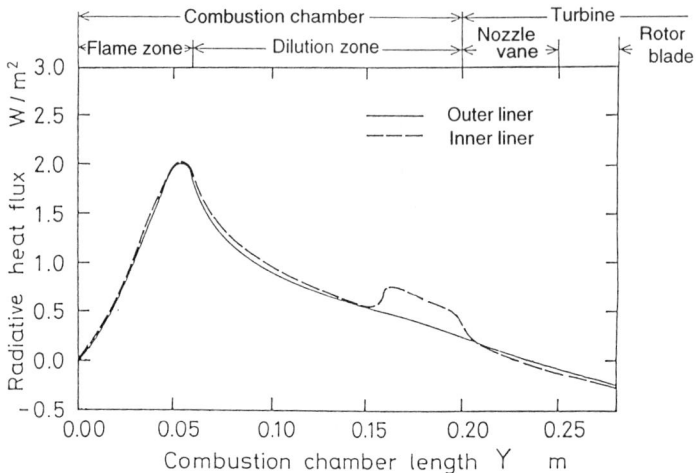

FIG. 7.25. Radiative heat flux along liners of combustion chamber (model 3)

distribution of radiative heat fluxes to the nozzle vane are presented in Fig. 7.26. When the gas absorptivity in the flame region is increased from 6 m^{-1} under the basic conditions to 40 m^{-1} under certain incomplete-combustion states, the radiative heat flux at the front end of the pressure

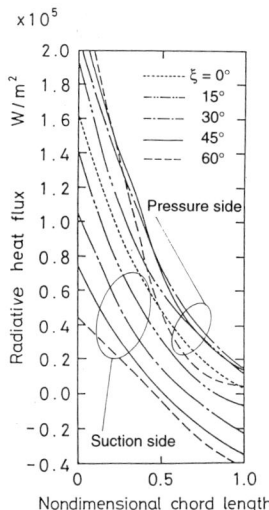

FIG. 7.26. Effects of stagger angle (model 1)

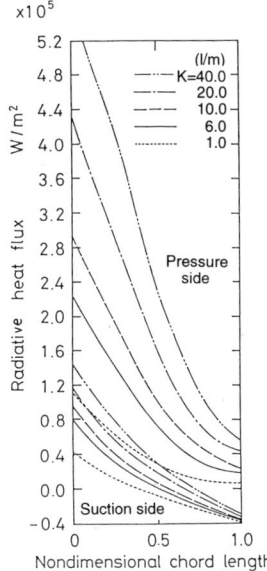

FIG. 7.27. Effects of flame absorption coefficient (model 1)

side increases by a factor of 2.5, as seen in Fig. 7.27. We know that the gas absorptivity can abruptly change due to a slight reduction in the air–fuel ratio in the presence of any soot formation in the flame, thus significantly affecting radiative heat flux. Figure 7.28 shows the effects of the emissivity of the nozzle vane surface on the radiative heat flux distribution. It is disclosed that the radiative heat flux at the front end of the pressure side is enhanced in direct proportion to an increase in the emissivity, but it decreases at the rear end of the suction side.

7.5. Nongray Gas (Combustion Gas) Layer

The gas consists of water vapor, carbon dioxide, and nitrogen, at the ratio of $H_2O : CO_2 : N_2 = 19.0 : 9.5 : 71.5$ mol %, which represents the combustion gas of methane [31, 32]. It is confined in infinite parallel black surfaces 1 and 2, at temperatures of $T_{w1} = 1{,}500$ K and $T_{w2} = 1000$ K, respectively. Figure 7.29 is a schematic of the physical system.

182 SOME INDUSTRIAL APPLICATIONS

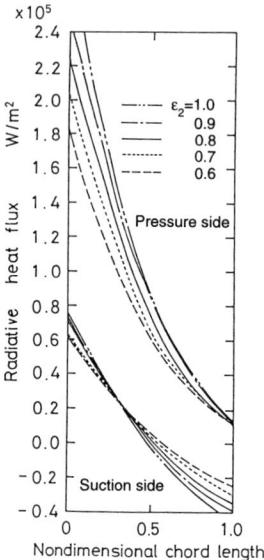

FIG. 7.28. Effects of nozzle vane emissivity (model 1)

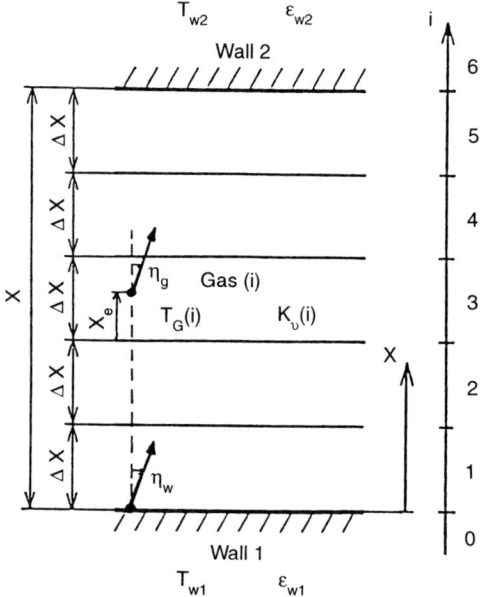

FIG. 7.29. Analytical model

7.5. NONGRAY GAS (COMBUSTION GAS) LAYER

The monochromatic absorption coefficient of each gas species is obtained by the following equation derived from Elsasser model [33, 34]:

$$K_\nu = (S/d) \frac{\sinh(2\pi r/d)}{\cosh(2\pi r/d) - \cos(2\pi \nu^*/d)}. \tag{7.4}$$

The ν^* is the deviation of the wave number from the nominal value of the absorption band. The values of the parameters in Eq. (7.4) are substituted for by the following parameters, α, ω, β, and Pe, obtained by using Edward's exponential wide-band model:

$$(S/d) = (\alpha/\omega) \exp(-\nu^*\omega), \tag{7.5}$$

$$(\pi r/d) = \beta \text{Pe}. \tag{7.6}$$

The substitution of Eqs. (7.5) and (7.6) into Eq. (7.4) results in the following equation, because the cosine term can be neglected:

$$K_\nu = (\alpha/\omega) \exp(-\nu^*\omega) \tanh(2\beta \text{Pe}). \tag{7.7}$$

Because the parameters α, ω, β, and Pe in Eq. (7.7) are dependent on temperature and pressure, the monochromatic absorption coefficient K_ν becomes a function of temperature, pressure, and wave number. To check the validity of parameters, emissivities of several gas volumes with uniform temperature are calculated using the parameters obtained from Edwards's exponential wide-band model [35]. Results agree well with the emissivity obtained from Hottel's chart.

Figure 7.30 is the analytical flowchart that is used to determine the temperature in each gas element and the wall heat flux of each wall element. Figure 7.31 compares the temperature profiles obtained by the present nongray analysis with those of gray analyses. It is seen that the gray analyses yield straight temperature profiles, whereas the temperature profiles of nongray analyses diminish rapidly near the higher temperature wall T_{w1} at $x/X = 0$ with its slope gradually leveling off toward the lower temperature wall T_{w2} at $x/X = 1.0$. These results reflect the typical difference in radiative heat transfer characteristics through gray and nongray gas layers.

Figure 7.32 depicts the spectra of radiative energy incident on the lower temperature wall from the higher temperature wall (dashed line) and from the whole gas layer (solid line), obtained by the nongray analysis. The dashed line shows that the radiative energy emitted from one wall (T_{w1}) transmits directly to the opposite wall (T_{w2}) through the "windows" within the wave number range. The radiative energy in the wave number range of the absorption bands of the nongray gas is almost completely absorbed. In contrast, the spectrum of the gas radiation exhibits its peaks at the wave number range of the absorption bands.

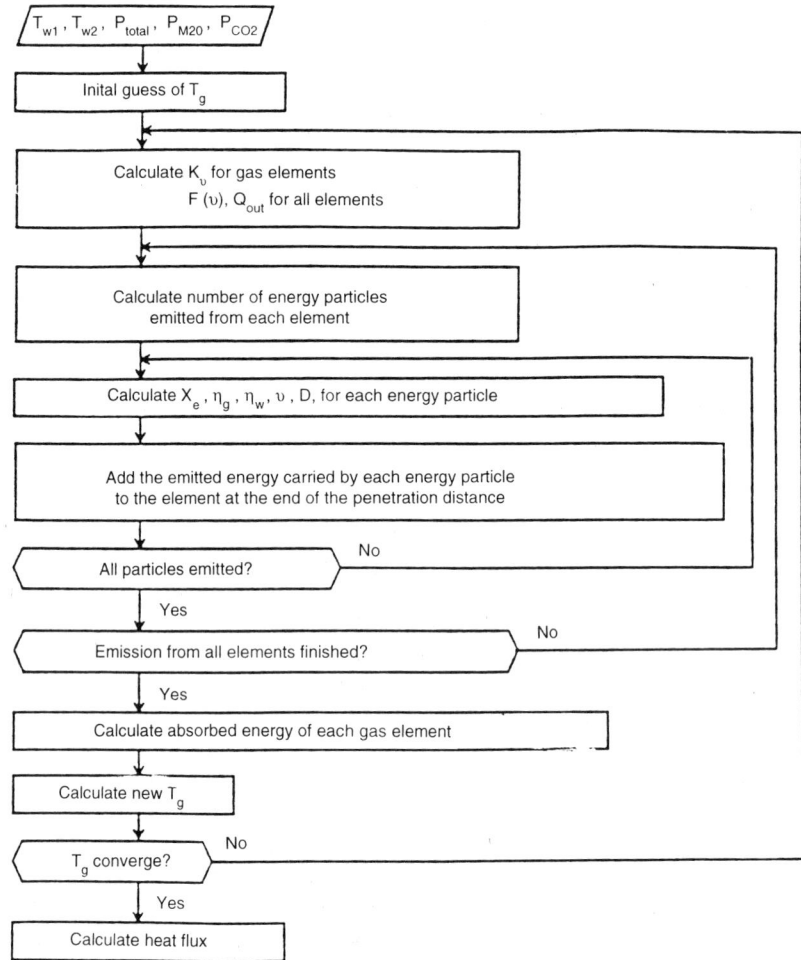

FIG. 7.30. Analytical flowchart

The net radiative heat flux incident on the lower temperature wall is listed in Table 7.3. Cases 1, 2, and 3 correspond to the cases in Fig. 7.31. It is seen that case 3, heat flux through the gray gas layer using an absorptivity obtained from Hottel's emissivity chart, is close to case 1, heat flux through the nongray layer. In contrast, case 2, heat flux through the gray gas layer using the Planck mean absorptivity, yields practically one-half of the heat flux through the nongray gas layer.

7.5. NONGRAY GAS (COMBUSTION GAS) LAYER

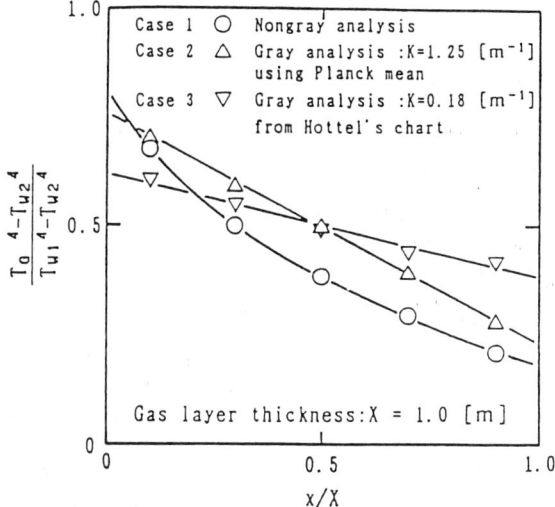

FIG. 7.31. Comparison of gray and nongray analyses

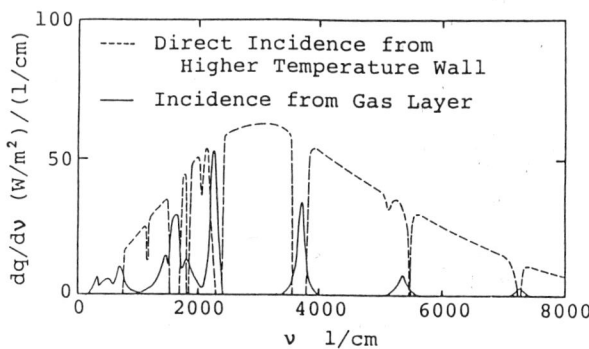

FIG. 7.32. Spectrum of radiative energy incident on the lower temperature wall obtained by nongray analysis

TABLE 7.3.

RADIATIVE HEAT FLUX INCIDENT ON THE LOWER TEMPERATURE WALL

	Case 1	Case 2	Case 3
Heat flux, kW/m^2	213.2	115.7	195.9

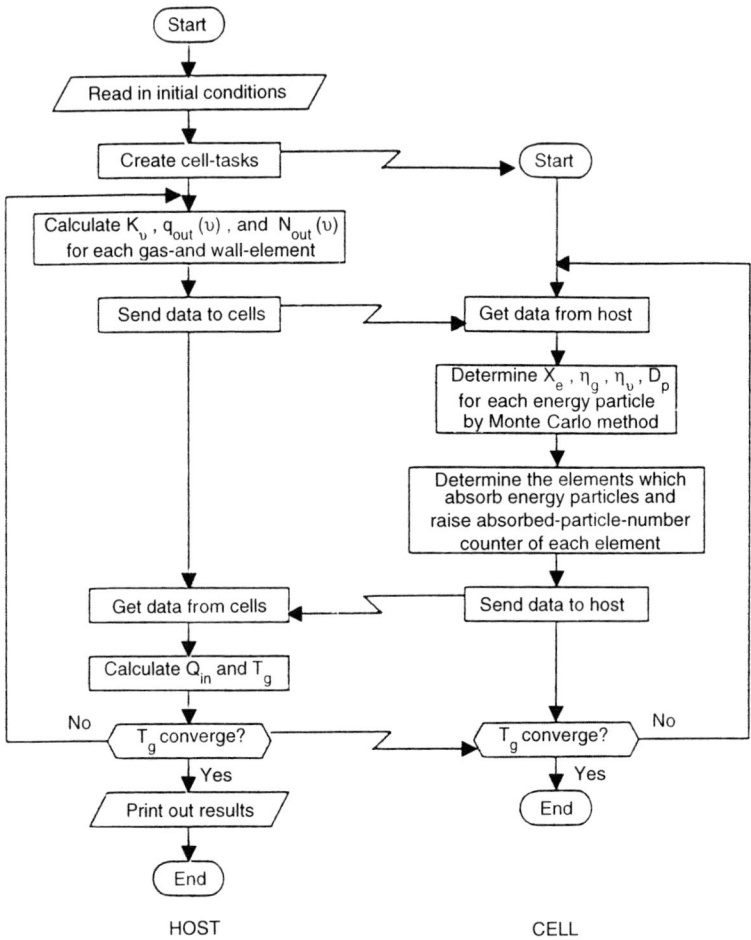

FIG. 7.33. Flowchart of parallel processing algorithm

The same problem ($T_{w1} = T_{w2} = 1000$ K; $x = 1$ m; $\varepsilon_1 = \varepsilon_2 = 1.0$; and $H_2O : CO_2 : N_2 = 0.1 : 0.1 : 0.8$ mol %) is solved using a Fujitsu experimental parallel computer AP 1000, in order to save computing time [32]. The computer consists of cell processors, ranging from 64 to 1024 in number. Figure 7.33 is a flowchart of the parallel processing algorithm for the radiative heat transfer analysis by the Monte Carlo method. The analysis is divided into three parts : (1) data input and determination of $N_{out}(\nu)$, number of energy particles emitted from each element; (2) seeking the loci of all energy particles by the Monte Carlo method; and

(3) determining the temperature of each gas element from the heat equation of each element. Parts (1) and (3) are carried out in the host computer; part (2) is performed in the cell system.

Table 7.4 shows how the computational time varies with the number of cells. It is observed that the computational time in the host computer is negligible compared with that consumed in the cell system. The computational time of a cell process is almost inversely proportional to the number of cells.

7.6. Circulating Fluidized Bed Boiler Furnace

The circulating fluidized bed boiler (CFBB) is characterized by a very high combustion efficiency for any solid fuels with very low fuel gas emission of SO_x and NO_x [36-38]. Combined radiation-convection heat transfer takes place among the gas, particles, and furnace walls. Figure 7.34 depicts a 2-D rectangular duct located above the secondary air inlet of a Studsvik 2.5-MW CFBB [39]. The system is 0.7×7.0 m in size with the walls at 600 K and an emissivity of 0.95. A gas with suspension particles at 1000 K enters through the lower wall and flows through the duct. The gas properties and vertical bulk densities are listed in Tables 7.5 and 7.6 [39], respectively. Note that Table 7.6 is derived from the experimental data in Fig. 7.35. Particle 1 refers to heat generating (coal) particles, and particle 2 represents non-heat-generating ones (bed materials). Their size and properties are given in Table 7.7. The bulk density ratio of particles 1 and 2 ranges from 1 to 49, as determined empirically. The gas velocity is

TABLE 7.4.
Variation of Computational Time with Cell Number

Number of cells	1	9	36	64
Communication (sec)				
Host → Cell	0.0576	0.0578	0.0577	0.0579
Cell → Host	0.0037	0.0656	0.2442	0.4071
Total	0.0613	0.1234	0.3019	0.4650
Calculation (sec)				
Host process	1.3148	1.7693	1.3522	1.3467
Cell process	9535.4164	1059.5728	264.8845	149.0074
Total	9536.7312	1061.2765	266.2367	150.3541
Total (sec)	9536.7925	1061.3999	266.5386	150.8191
Speedup ratio	1.0	8.985	35.78	63.23
Efficiency	(100%)	99.8%	99.4%	98.8%

188   SOME INDUSTRIAL APPLICATIONS

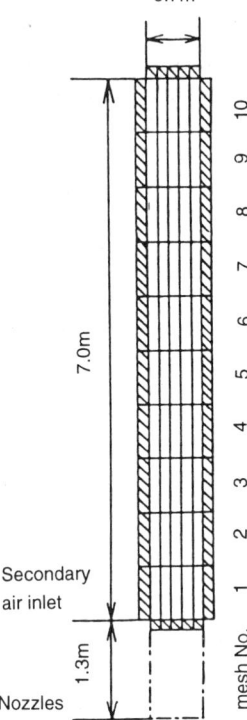

FIG. 7.34. Theoretical model

uniformly upward at 4.8 m/sec and the mass flux of particles is 12 kg/m² sec⁻¹. The upper and lower walls are porous and black at 1000 and 600 K, respectively.

The following heat transfer actions are taken into account in the model:

TABLE 7.5.
SOME PROPERTIES OF GAS

Absorption coefficient (m^{-1})	$a_8 = 0.2$
Density (kg/m^3)	$\rho_9 = 0.32$
Special heat (J/kg K^{-1})	$c_{p8} = 1160$
Thermal conductivity (W/m K^{-1})	$\lambda_8 = 0.0717$
Prandtl number	Pr $= 0.742$
Kinematic viscosity coefficient (m^2/sec)	$\nu = 1.43 \times 10^{-4}$

7.6. CIRCULATING FLUIDIZED BED BOILER FURNACE

TABLE 7.6.
BULK DENSITY (kg/m³)

Mesh	Particle 1	Particle 2	Total
10	0.2	9.8	10.0
9	0.24	11.76	12.0
8	0.3	14.7	15.0
7	0.34	16.66	17.0
6	0.4	19.6	20.0
5	0.5	24.5	25.0
4	0.6	29.4	30.0
3	0.8	39.2	40.0
2	1.2	58.8	60.0
1	1.8	88.2	90.0

1. Radiation from heat-generating coal particles in combustion, non-heat-generating particles of bed materials, gas, and enclosure wall
2. Absorption of radiation
3. Anisotropic scattering by particles
4. Heat released by heat-generating particles
5. Convective heat transfer between gas and particles
6. Convective heat transfer from the bed to the wall.

The radiative characteristics of the gas with particles are determined by the shape, size, optical characteristics (emissivity) of the particles, particle

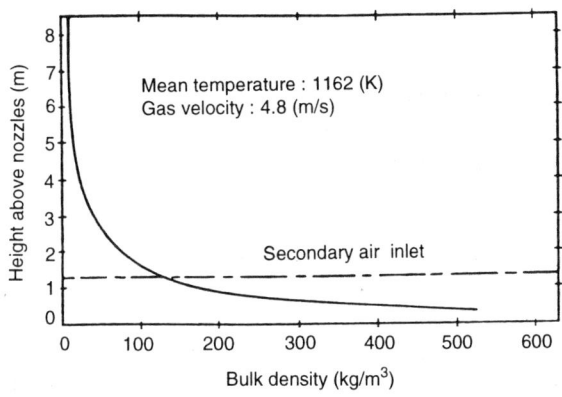

FIG. 7.35. Experimental results of vertical bulk density

TABLE 7.7.
SIZE AND SOME PROPERTIES OF PARTICLES

Diameter (μm)	$d_{s1} = 240$	$d_{s2} = 240$
Emissivity	$\varepsilon_{s1} = 0.85$	$\varepsilon_{s2} = 0.6$
Density (kg/m^3)	$\rho_{s1} = 1300$	$\rho_{s2} = 3000$
Specific heat (J/kg K^{-1})	$c_{0s1} = 1000$	$c_{0s2} = 1300$

number density (N_{s1}, N_{s2}), and the thermal radiative characteristics of the gas and particles and the scattering coefficient. The ω denotes the scattering albedo. Then, one has

$$\beta = a_g \psi + a_{s1} + a_{s2} + \sigma_{s1} + \sigma_{s2}, \quad (7.8)$$

$$\omega = (\sigma_{s1} + \sigma_{s2})/\beta, \quad (7.9)$$

where

ψ = void fraction

= volume of gas / volume of gas and particles,

$1 - \psi$ = volume of particles / volume of gas and particles

$$= \frac{4}{3}\pi(d_{s1}/2)^3 N_{s1} + \frac{4}{3}\pi(d_{s2}/2)^3 N_{s2}, \quad (7.10)$$

$$a_{s1} = \varepsilon_{s1}\pi(d_{s1}/2)^2 N_{s1}, \quad (7.11)$$

$$a_{s2} = \varepsilon_{s2}\pi(d_{s2}/2)^2 N_{s2}, \quad (7.12)$$

$$\sigma_{s1} = (1 - \varepsilon_{s1})\pi(d_{s1}/2)^2 N_{s1}, \quad (7.13)$$

$$\sigma_{s2} = (1 - \varepsilon_{s2})\pi(d_{s2}/2)^2 N_{s2}, \quad (7.14)$$

and a and σ are absorption and scattering, respectively. The subscripts s1, s2, and g denote particles 1, 2, and gas, respectively. These particles have the anisotropic phase function

$$\phi(\eta) = 8/(3\pi)(\sin\eta - \eta\cos\eta), \quad (7.15)$$

which determines the direction of scattering energy. It is valid for a sphere with diffuse surface and strong backward scattering characteristics. Here, the particles are treated optically, since their size (240 μm) exceeds 20 times the wavelength range (under 10 μm) of radiation by a blackbody at 1000 to 1200 K. These equations are solved by a Newton-Raphson method numerically, to determine the temperatures of the gas and particles as well

7.6. CIRCULATING FLUIDIZED BED BOILER FURNACE

as the wall heat flux distribution. The energy equations for the gas, particles, and wall elements are

gas:
$$4\sigma a_g \psi T_g^4 \Delta V + Q_{cgw} + Q_{cgs1} + Q_{cgs2} + Q_{cgs2} + Q_{f,\text{out},g}$$
$$= Q_{r,\text{in},g} + Q_{f,\text{in},g}, \tag{7.16}$$

particle 1:
$$4\sigma a_{s1} T_{s1}^4 \Delta V + Q_{cs1w} + Q_{f,\text{out},s1}$$
$$= Q_{r,\text{in},s1} + Q_{cgs1} + Q_{hs1} + Q_{f,\text{in},s1}, \tag{7.17}$$

particle 2:
$$4\sigma a_{s2} T_{s2}^4 \Delta V + Q_{cs2w} + Q_{f,\text{out},s2}$$
$$= Q_{r,\text{in},s2} + Q_{cgs2} + Q_{f,\text{in},s2}, \tag{7.18}$$

wall:
$$\varepsilon_w \sigma T_w^4 \Delta S + Q_a = Q_{r,\text{in},w} + Q_{cgw} + Q_{cs1w} + Q_{cs2w}. \tag{7.19}$$

Here, $Q_{r,\text{out}}$ is the total radiative energy emitted from an element and absorbed by other elements; ΔV, element volume of gas; ΔS, element area of wall; σ, Stefan-Boltzmann constant, Q_a, heat load on wall; Q_f, enthalpy transport; Q_{cgw}, Q_{cs1w}, Q_{cs2w}, convective heat transfer between the wall and the gas, between the wall and particle 1, between the wall and particle 2, respectively; and Q_{cgs1}, Q_{cgs2}, convective heat transfer between the gas and particles 1 and 2, respectively. The term Q_{hs1} is the heat generation rate by particle 1, assumed to be proportional to the bulk density. The total heat generation is 2.5 MW. The value of $Q_{r,\text{in}}$ can be obtained by

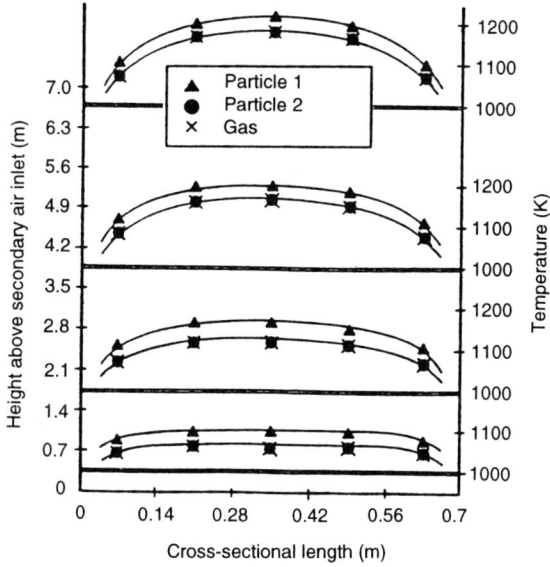

FIG. 7.36. Theoretical results of temperature distribution

adding all the energy components transferred from all other gases and wall elements by means of the READ method. Convective heat transfer from the gas and particles to the furnace walls is determined using Martin's model [40].

Results are presented in Figs. 7.36, 7.37, and 7.38 for the gas and particle temperature distribution, wall heat flux distribution, and wall-gas heat transfer coefficient, respectively. Figure 7.36 shows both the gas and particle temperatures are in the range of 1000 to 1200 K, which coincides with the experimental mean temperature of 1162 K in the furnace. The gas temperature is about the same as that of the non-heat-generating particles but is lower by about 30°C than that of the heat-generating particles. Figure 7.37 reveals that in the lower part of the bed with higher bulk density, the radiative heat flux is about the same order of magnitude as the convective heat flux. The radiative heat flux dominates in the upper part. Theoretical and experimental results of the wall heat transfer coefficient agree well qualitatively, as seen in Fig. 7.38.

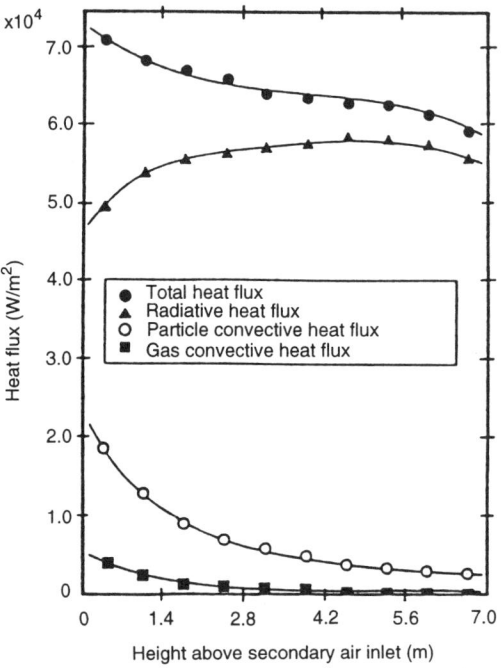

FIG. 7.37. Theoretical results of wall heat flux

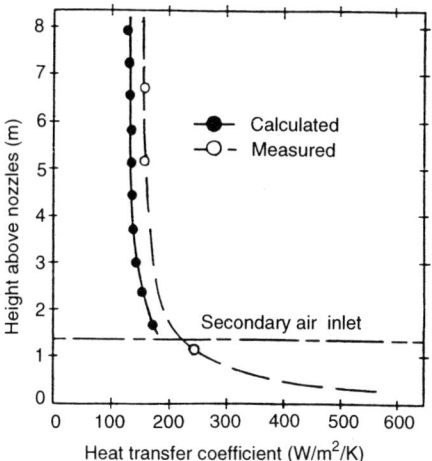

FIG. 7.38. Theoretical and experimental results of wall-gas heat transfer coefficient

Instead of deriving the vertical bulk density distribution from the experimental data in Fig. 7.35, it is theoretically determined by numerically solving one-dimensional, gas-particle, two-phase flow equations, a set of seven equations with seven unknowns [37]. Theoretical results of the bulk density distribution agree well with measured data. The study [37] concludes that the horizontal density distribution of particles should be taken into account in the analysis to improve the accuracy. The effects of particle size on the bulk density, wall heat flux, and gas and particle temperatures are determined. Taniguchi et al. [38] employed two computer programs: one to determine the vertical bulk density distribution in the furnace by solving one-dimensional, gas-particle two-phase flow equations, and the other to determine thermal behavior in a two-dimensional CFBB furnace. The effects of system geometry and size on thermal behavior in the furnace are investigated.

7.7. Three-Dimensional Systems

Two 3-D systems are presented in this section.

7.7.1. RADIATION IN 3-D PACKED SPHERES

An optical experiment reveals the importance of side effects on the transmittance of radiative energy through 3-D packed spheres [41]. To

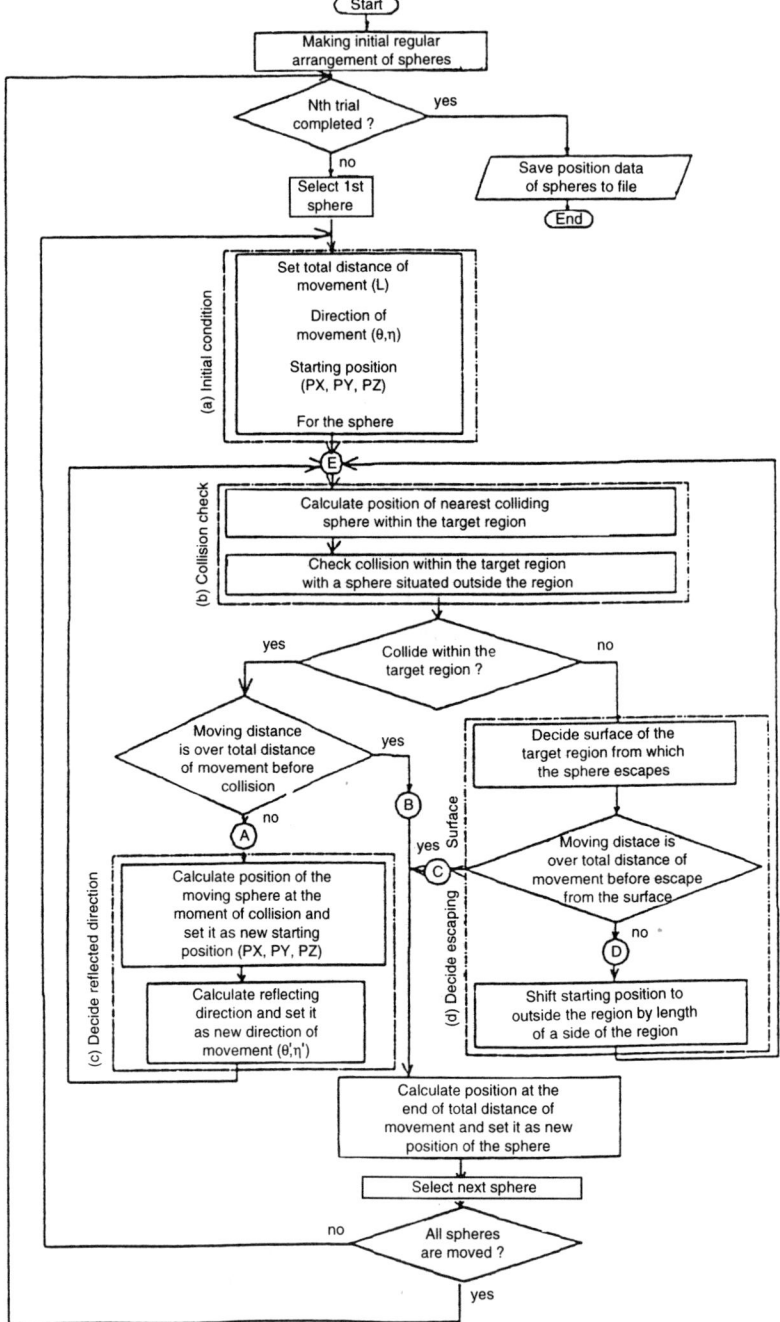

FIG. 7.39. Computational procedure to obtain randomly packed spheres

7.7. THREE-DIMENSIONAL SYSTEMS

analyze numerically the radiative transmittance through a vessel filled with randomly packed, equal-diameter spheres, a computer program is developed to generate 3-D randomly packed spheres using the flowchart of Fig. 7.39. The radiative heat transfer through this bed is analyzed by means of the Monte Carlo method. Theoretical results of transmittance compare well with test data. Figure 7.40 shows the effects of regularity of the sphere arrangement on the radiative transmittance through packed spheres. Model a refers to the periodic boundary case in which the striking energy bundle is reentered from the opposite side wall in the direction parallel to the one before collision. It simulates the packed spheres of infinite width. The figure indicates that τ of regularly packed spheres is lower than that of randomly packed ones except for α (absorption coefficient of sphere surface) = 1.0.

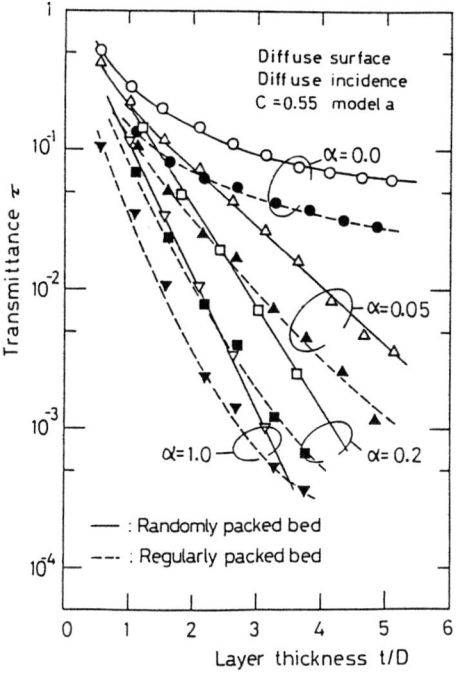

FIG. 7.40. Effects of regularity of sphere arrangement on radiative transmittance through packed spheres

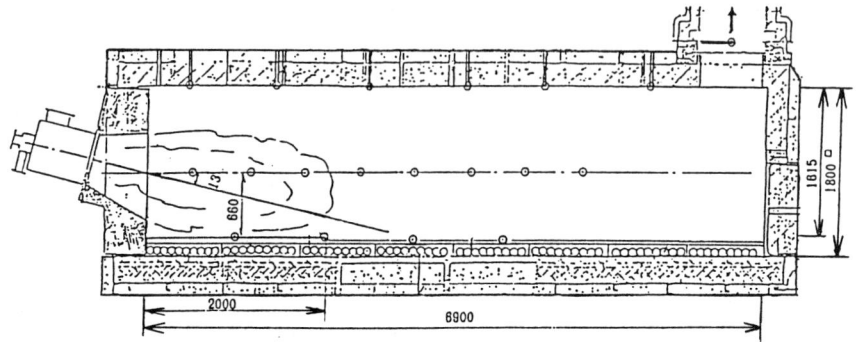

Fig. 7.41. Dimensions of test furnace

7.7.2. 3-D Industrial Gas-Fired Furnaces

Figure 7.41 is a schematic of an industrial gas-fired furnace [42]. The grid networks on the *x-z* and *x-y* planes for numerical computations are presented in Fig. 7.42. Radiative heat transfer is treated by the READ method, turbulent conventive heat transfer is determined with the aid of the k-ε two-equation turbulent flow model, and conductive heat transfer in the furnace walls is taken into account. The energy balance equations are solved in terms of unknown temperatures using the Newton-Raphson method.

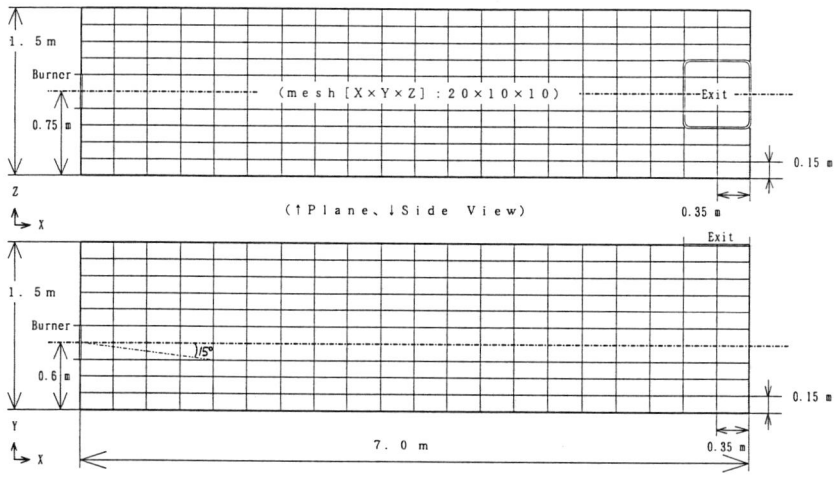

Fig. 7.42. Grid network

7.7. THREE-DIMENSIONAL SYSTEMS

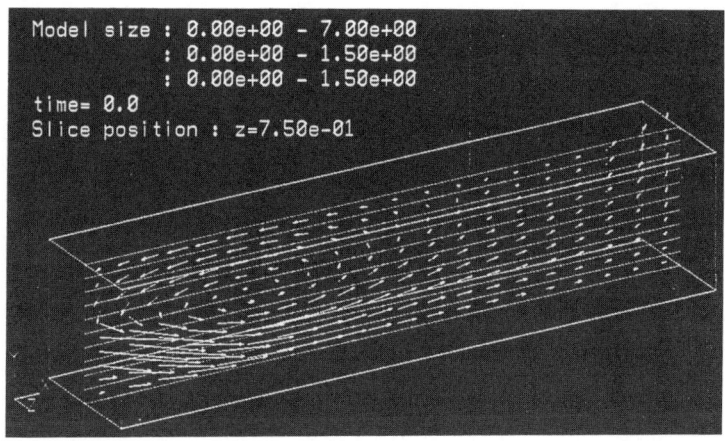

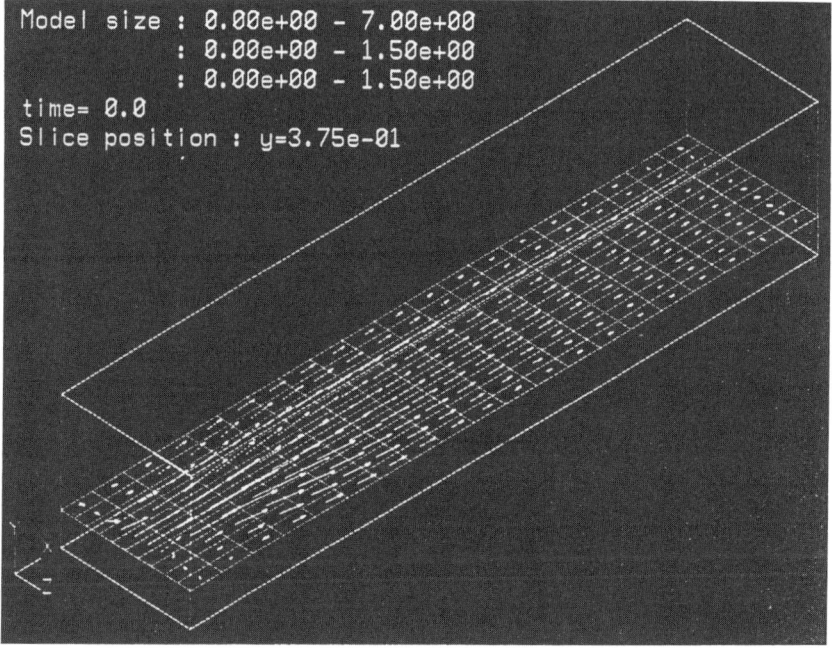

FIG. 7.43. Velocity vector distribution in the furnace (cross-section along the vertical axis: Z = 5)

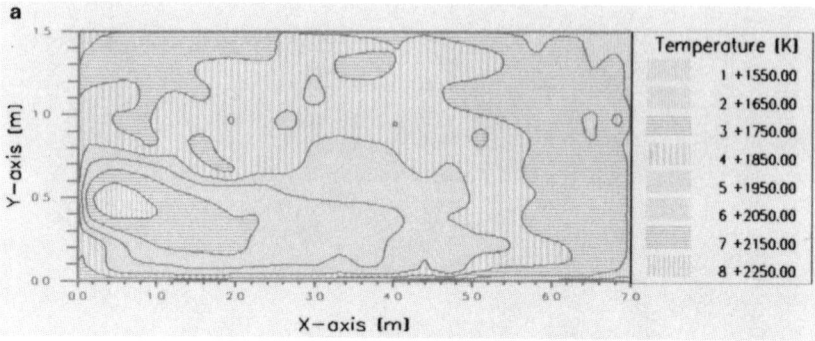

FIG. 7.44. (a) Temperature distribution in the furnace (cross-section along the vertical axis: Z = 5)

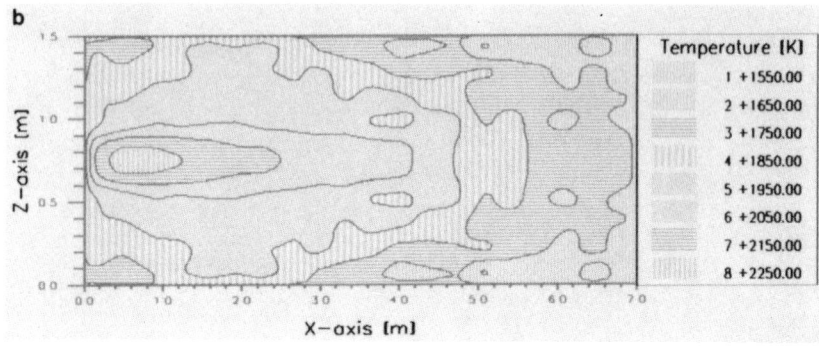

FIG. 7.44. (b) Temperature distribution in the furnace (cross-section along the horizontal axis: Y = 3)

A numerical analysis is performed on a test furnace of 1.5-m width, 7.0-m length, and 1.5-m height, using a luminous-flame natural-gas-fired burner. The firing rate is 3.50×10^6 kcal/hour (approximately 4 MW). Air enters the furnace at 20°C and 15 m/sec through a 300- × 300-mm square inlet port with 10% excess air ratio. The emissivity of the furnace wall

7.7. THREE-DIMENSIONAL SYSTEMS

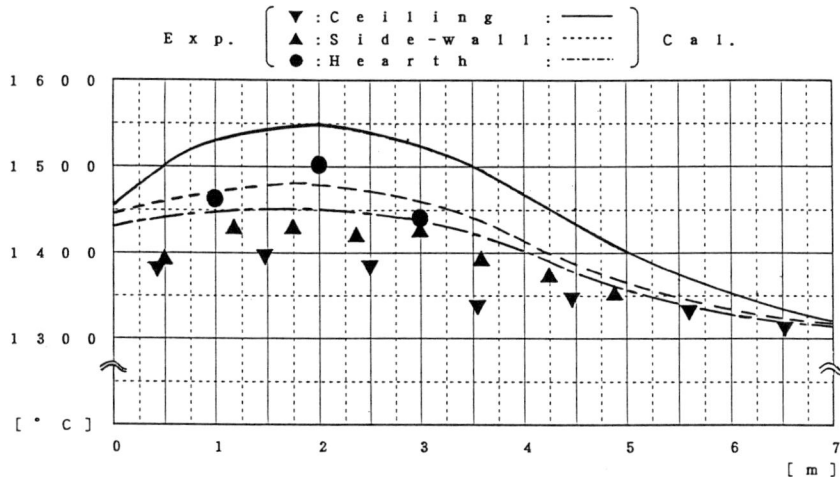

FIG. 7.45. Temperature distribution on the furnace walls

surfaces is 0.9. The inlet and outlet ports are porous and black. The gas absorption coefficient is 0.5 for the flame region and 0.2 for the flue gas.

Results are presented in Figs 7.43, 7.44, and 7.45 for the distributions of velocity vectors in the furnace, temperature in the furnace, and temperature in the furnace walls, respectively.

References

1. Taniguchi, H. (1967). Temperature distribution of radiant gas calculated by Monte Carlo method. *Bull. JSME* **10**(42), 975–988.
2. Taniguchi, H. (1969). The radiative heat transfer of gas in a three dimensional system calculated by Monte Carlo method. *Bull. JSME* **12**(49), 67–78.
3. Taniguchi, H., and Funazu, M. (1970). The numerical analysis of temperature distributions in a three dimensional furnace. *Bull. JSME* **13**(66), 1458–1468.
4. Taniguchi, H., Sugiyama, K., and Taniguchi, K. (1974). The numerical analysis of temperature distribution in a three dimensional furnace (2nd Report: The comparison with experimental results). *Heat Transfer—Jpn. Res.*, **3**(4), 41–54.
5. Kobiyama, M., Taniguchi, H., and Saito, T. (1979). The numerical analysis of heat transfer combined with radiation and convection (1st Report: The effect of two-dimensional radiative transfer between isothermal parallel plates). *Bull. JSME* **22**(167), 707–714.
6. Taniguchi, H., Yang, W.-J., Kudo, K., Hayasaka, H., Oguma, M., Kusama, A., Nakamachi, I., and Okigami, N. (1986). Radiant transfer in gas filled enclosures by radiant energy absorption distribution method. *Heat Transfer, Proc. Int. Heat Transfer Conf.*, 8th, San Francisco, 1986, Vol. 2, pp. 757–762.
7. Taniguchi, H., Kudo, K., and Yang, W.-J. (1988). Advances in computational heat transfer by Monte Carlo method. *Comput. Mech. '88, Theory Appl., Proc. Int. Conf. Comput. Eng. Sci.*, Atlanta, GA, 1988, Vol. 2, pp. 56ii1–56ii4.
8. Nakamura, T., Omori, T., Yasuzawa, K., Nakamachi, I., and Taniguchi, H. (1987). Radiative heat transfer analysis in a forge furnace. *Numer. Methods Therm. Prob., Proc. Int. Conf.*, 5th, Montreal, Canada, 1987, Vol. V, Part 1, pp. 845–856.
9. Obata, M., Funazaki, K., Taniguchi, H., Kudo, K., and Kawaski, M. (1989). Numerical simulation of radiative transfer to high pressure turbine nozzle vanes of aero-engines. *Numer. Methods Therm. Prob., Proc. Int. Conf.*, 6th, Swansea, U.K., 1989, Vol. VI, Part 1, pp. 751–761.
10. Omori, T., Taniguchi, H., and Kudo, K. (1989). Radiative heat transfer analysis of indoor thermal environment. *Numer. Methods Therm. Prob., Proc. Int. Conf.*, 6th, Swansea, U.K., 1989, Vol. VI, Part 1, pp. 730–740.
11. Taniguchi, H., Yang, W.-J., Kudo, K., Hayasaka, H., Fukuchi, T., and Nakamachi, I. (1988). Monte Carlo method for radiative heat transfer analysis of general gas-particle enclosures. *Int. J. Numer. Methods Eng.* **25**(2), 581–592.
12. Mengüç, M. P., and Viskanta, R. (1983). Comparison of radiative transfer approximations for a highly forward scattering planar medium. *J. Quan. Spectros. Radiat. Transfer* **29**(5), 381–394.
13. Kudo, K., Taniguchi, H., and Fukuchi, T. (1988). Radiative heat transfer analysis in emitting-absorbing-scattering media by the Monte Carlo method (anisotropic scattering effects). *Heat Transfer—Jpn. Res.* **17**(2), 87–97.
14. Taniguchi, H., Kudo, K., Katayama, T., Nakamura, T., Fukuchi, T., and Kumagai, N. (1987). Scattering effects of coupled convection-radiation heat transfer of multi-dimension in gas flow with fine particles, in "Coal Combustion, Science and Technology of Industrial and Utility Applications" (Proc. Int. Symp. Coal Combust., Beijing, China), pp. 299–306. Hemisphere Publishing, Washington, DC.
15. Kudo, K., Taniguchi, H., Kaneda, H., Yang, W.-J., Zhang, Y.-Z., Guo, K.-H., and Matsumura, M. (1990). Flow and heat transfer simulation in circulating fluidized beds. *Prep. Int. Conf. Circ. Fluid. Beds*, 3rd, Nagoya, Japan, pp. 5.8.1–5.8.6.
16. Tien, C. L. (1988). Thermal radiation in packed and fluidized beds. *Heat Transfer* **110**(B), 1230–1242.

17. Taniguchi, H., Kudo, K., Yang, W.-J., and Kim, Y.-M. (1989). Numerical analysis on transmittance of radiative energy through three-dimensional packed spheres. *Numer. Methods Therm. Probl., Proc. Int. Conf.*, 6th, Swansea, U.K., 1989, Vol. VI, Part 1, pp. 762–772.
18. Kudo, K., Tanigchi, H., Kim, Y.-M., and Yang, W.-J. (1991). Transmittance of radiative energy through three-dimensional packed spheres. *Proc. ASME-JSME Therm. Eng. Jt. Conf.*, 3rd, Reno, NV, ASME Book No. 10309D, pp. 35–42.
19. Taniguchi, H., Kudo, K., Otaka, M., and Sumarusono, M. (1991). Non-gray analysis on radiative energy transmittance through real gas layer by Monte Carlo method. *Numer. Methods Therm. Probl., Proc. Int. Conf.*, 7th, Stanford, CA, 1991, Vol. VII, Part 1, pp. 748–757.
20. Hottel, H. C. (1954). Radiant-heat transmission. In "Heat Transmission" (W. McAdams, ed.), 3rd ed., pp. 82–89. McGraw-Hill, New York.
21. Siegel, R., and Howell, J. R. (1981). "Thermal Radiation Heat Transfer," 2nd ed., pp. 522–523. Hemisphere Publishing, Washington, DC.
22. Heaslet, M. A., and Robert, F. W. (1965). Radiative transfer and wall temperature slip in an absorbing planar medium. *Int. J. Heat Mass Transfer*, **8**(7), 979–994.
23. Hottel, H. C., and Cohen, E. S. (1958). Radiant heat exchange in a gas-filled enclosure (allowance for nonuniformity of gas temperature). *AIChE J.*, **4**(1), 3–14.
24. Howell, J. R. (1968). Application of Monte Carlo to heat transfer problems. *Adv. Heat Transfer* **5**, 1–54.
25. Hottel, H. C., and Sarofim, A. F. (1967). "Radiative Transfer," Chapters 7–8. McGraw-Hill, New York.
26. Ross, S. M. (1985). "Introduction to Probability Models," 3rd ed., pp. 437–438. Academic Press, Orlando, FL.
27. Kunitomo, T., Matsuoka, K., and Oguri, T. (1975). Prediction of radiative heat flux in a diesel engine. *SAE Trans.* **84**, 1908–1917.
28. Taniguchi, H., Yang, W.-J., Kudo, K., Hayasaka, H., and Fukuchi, T. (1988). Monte Carlo method for radiative heat transfer analysis of general gas-particle enclosures. *Int. J. Numer. Methods Eng.*, **25**, 581–592.
29. Kudo, K., Taniguchi, H., and Fukuchi, T. (1989). Radiative heat transfer analysis in emitting-absorbing-scattering media by the Monte Carlo method (anisotropic scattering effects). *Heat Transfer—Jpn. Res.*, **18**, 87–97.
30. Kudo, K., Taniguchi, H., Guo, K.-F., Katayama, T., and Nagata, T. (1991). Heat transfer simulation in a furnace for steam reformer. *Kagaku Kogaku Ronbunshu* **17**(1), 103–110.
31. Taniguchi, H., Kudo, K., Otaka, M., Sumarsono, M., and Obata, M. (1991). Non-gray analysis of radiative energy transfer through real gas layer by Monte Carlo method. *Numer. Methods Therm. Probl., Proc. Int. Conf.*, 7th, Stanford, CA, 1991, Vol. VII, Part 1, pp. 748–757.
32. Taniguchi, H., Kudo, K., Ohtaka, M., Mochida, A., Komatsu, T., Kosaka, S., and Fujisaki, M. (1992). Monte Carlo simulation of non-gray radiation heat transfer on highly parallel computer AP1000. In "Transport Phenomena Science and Technology" (B.-X. Wang, ed.), pp. 581–586, Higher Education Press, Beijing.
33. Edwards, D. K. (1967). Radiation heat transfer in nonisothermal nongray gases. *J. Heat Transfer* **89**(3), 219–229.
34. Edwards, D. K. (1981). "Radiation Heat Transfer Notes," p. 195. Hemisphere Publishing, Washington, DC.
35. Edwards, D. K. et al. (1973). Thermal radiation by combustion cases. *Int. J. Heat Mass Transfer* **16**, 25–40.

36. Kudo, K., Yang, W.-J., Taniguchi, H., Kaneda, H., Matsumura, M., and Guo, K.-H. (1990). Combined radiation-convection heat transfer analysis in a circulating fluidized bed boiler. *Bull. Fac. Eng.*, Hokkaido Univ. **150**, 17–23.
37. Kudo, K., Taniguchi, H., Kaneda, H., Yang, W.-J., Zhang, Y. Z., Guo, K.-H., and Matsumura, M. (1990). Flow and heat transfer simulation in circulating fluidized beds. In "Circulating Fluidized Bed Technology III" (P. Basu, M. Horio, and M. Hasatani, eds.), pp. 269–274, Pergamon, Oxford.
38. Taniguchi, H., Yang, W.-J., Kudo, K., Wang, Y., Guo, K.-H., Matsumura, M., and Kaneda, H. (1991). Radiative heat transfer in a circulating fluidized-bed boiler furnace by a Monte Carlo method. In "Proceedings of the Second European Conference on Industrial Furnaces and Boilers" (A. Rels, J. Ward, R. Collin, and W. Leuckel, eds.), Vol. II, pp. II-14-1 to II-14-9, Vilamoura, Algare, Portugal.
39. Johnsson, F., Andersson, B. A., and Leckner, B. (1984). Heat transfer in FBB. *IEA Meet. Math. Model.*, Boston, pp. 1–18.
40. Martin, H. (1984). Heat transfer between gas fluidized beds of solid particles and the surfaces of immersed heat exchanger elements. *Chem. Eng. Prog.*, **18**, 157–233.
41. Kudo, K., Taniguchi, H., Kim, Y.-M., and Yang, W.-J. (1991). Transmittance of radiative energy through three-dimensional packed spheres. *Proc. ASME-JSME Therm. Eng. Jt. Conf.*, 2nd, Vol. 4, pp. 35–42.
42. Matsumurra, M., Ito, S., Ichiraku, Y., and Saeki, T. (1992). Heat transfer simulation in industrial gas furnaces. *Proc. Int. Gas Res. Conf.*, Industrial Utilization, Vol. 5, pp. 244–253.

APPLICATIONS ON DISK

The following programs used in this book are included in the floppy disk as text files. The figure numbers following the file names correspond with the figures in the book.

TFM.FOR	(Fig. 2.4)
ZM. FOR	(Fig. 2.13)
BEER.FOR	(Fig. 3.4)
RAT1.FOR	(Fig. 3.16)
RAT2.FOR	(Fig. 3.23)
RADIAN.FOR	(Fig. 6.2)
RADIANW.FOR	(Fig. 6.11)
RADIAN1.FOR	(Fig. 7.3)

They are copied to the disk in the 2HD 1.44MB form for IBM-PCs or compatibles. They are source files written by Fortran 77 and can be compiled and linked when using an appropriate Fortran compiler. They are successfully compiled, linked, and executed by using Microsoft Fortran Version 5.1 for the MS-DOS operating system. Please make your own backup disk by using the COPY or DISK COPY command and use that disk for further access, in order to protect your original disk from unfortunate file destruction.

All the programs require a printer to output the results. An appropriate printer, which can be recognized by your DOS as the DOS device name PRN should be connected to your computer. To check whether your printer has the correct setting, simply do the following:

1. Run DOS
2. Insert the floppy disk into drive A: of your computer
3. Key in the following instruction
. PRINT A:BEER.FOR
4. Press the return key twice.

When the list from the file A: BEER.FOR begins to print, then your printer is appropriately set. If it does not print, then consult the instruction manual.

LIST OF IMPORTANT VARIABLES IN COMPUTER PROGRAMS

ABSORP	m^{-1}	calculated absorption rate of a unit length
ABSPR	m^{-1}	exact value of absorption rate of a unit length
AKD		optical thickness of gas layer
AK (NG)	m^{-1}	gas absorption coefficient
AL		direction cosine of energy particle locus along X axis
AM		direction cosine of energy particle locus along Y axis
ANEWG (NG)		coefficient of equation for TG of gas element
ANEWW (NW)		coefficient of equation for TW of wall element
ANRAY		real function of NRAY
AS		self-absorption ratio
ASG (NG)		self-absorption ratio of gas element
ASW (NW)		self-absorption ratio of wall element
AVHG	kW/m^3	average heat load in furnace
BNEWG (NG)		coefficient of equation for TG of gas element
BNEWW (NW)		coefficient of equation for TW of wall element
CNEWG		coefficient of equation for TG of gas element
CNEWW		coefficient of equation for TW of wall element
CP (NG)	$J/kg\ K^{-1}$	specific heat of gas
CPO	$J/kg\ K^{-1}$	specific heat of inlet gas
DAK		optical thickness of a gas element
DELTAT		correction of T by Newton-Raphson method
DLW (NW)	m	length of wall element
DXG	m	width of gas element
DYG	m	height of gas element
E3 (X)		third exponential integral function
EM (NW)		wall emissivity
EM1, EM2		wall emissivity
ERR		maximum relative correction of TGs and TWs

LIST OF IMPORTANT VARIABLES IN COMPUTER PROGRAMS 205

ERRG		relative correction of TG for each gas element
ERRN		error in iteration of Newton-Raphson method
ERRW		relative correction of TW for each wall element
ETA	rad	polar angle
ETAG	rad	polar angle
ETAW	rad	polar angle
GG	m^2	total exchange area
GM	$kg/m^2\ sec^{-1}$	incoming mass flow
GMF (IW, NG)	$kg/m^2\ sec^{-1}$	incoming mass flow through IWth boundary of gas element NG
GMF1 (IW, NG)	$kg/m^2\ sec^{-1}$	incoming mass flow through IWth boundary of gas element NG
GMF2 (IW, NG)	$kg/m^2\ sec^{-1}$	incoming mass flow through IWth boundary of gas element NG
GP (NG)		data of gas elements transferred to subroutine PRTDAT
H (NW)	$W/m^2\ K^{-1}$	heat transfer coefficient
INDABS		index of adsorption, 1 : absorbed, 0 : transmitted or reflected
INDFL		1 : luminous flame, 0 : nonluminous flame
INDFUL		1 : full load, 0 : half load
INDGW (NG)		1 : gas element, 0 : out of the system
INDGWC		index of next element, 1 : gas, 0 : wall
INDNT1 (IW, NG)		number of element next to IWth boundary, > 0 : wall, < 0 : gas
INDNT2 (IW, NG)		number of element next to IWth boundary, > 0 : wall, < 0 : gas
INDXT (IW, NG)		number of element next to IWth boundary, > 0 : wall, < 0 : gas
INDRDP		1 : print READ values, 0 : supress printing READ values
INDWBC (NW)		boundary condition of wall element 1 : temperature given, 0 : heat flux given
INRAY		number of energy particles already emitted
IUP		number of upstream gas elements
IW		index of gas element boundary, 1 : left, 2 : top, 3 : right, 4 : bottom

IWMAX	maximum value of IW
IX	number of emitting element in X coordinate
IXA	number of absorbing element in X coordinate
IXT	number of target element in X coordinate
IY	number of emitting element in Y coordinate
IYA	number of absorbing element in Y coordinate
IYT	number of target element in Y coordinate
KA	1: emitted from wall, 0: emitted from gas
N	number of division of gas layer
ndisp	index used to display the number of emitted particles
NG	number of gas elements
NG2	(= NG − 20)
NGE	number of gas elements where energy particle exits at the time
NGET	number of next element
NGIN	number of gas elements in the system
NGM	maximum value of NG
NGMAX	maximum value of NG
NGS	number of source gas elements in QRIN calculation
NRAY	number of energy particles emitted from a gas element
NRD	number of absorbed energy particles
NW	number of wall elements
NWE	number of wall element on which energy particle collide
NWH	number of next wall element
NWM	maximum value of NW
NWMAX	maximum value of NW
NWS	number of source-wall elements in QRIN calculation
OUTRAY	energy particle numbers absorbed outside the emitting element
PAI	the circular constant

LIST OF IMPORTANT VARIABLES IN COMPUTER PROGRAMS 207

QG (NG)	W/m^3	heat load in gas element
QND		nondimensional wall heat flux
QRING	W	absorbed radiative energy by gas element
QRINW	W	absorbed radiative energy by wall element
QW (NW)	W/m^2	net wall heat flux
RAN		uniform random number
RAND		seed of random number
RD		READ value
RDGG (NG, NG)		READ value between gas-gas elements
RDGW (NG, NW)		READ value between gas-wall elements
RDWG (NW, NG)		READ value between wall-gas elements
RDWW (NW, NW)		READ value between wall-wall elements
S	m	traveling length of energy particle
S (IW)	m	variable used for obtaining pass length within a gas element
S1S1, S1S2		total exchange area between bounding walls
SG		total exchange area between wall and gas elements
SBC	W/m^2 K^{-4}	Stefan-Boltzmann constant
SMIN	m	pass length within a gas element (minimum of positive S(IW)s)
SW (NW)	m^2	area of wall element
TG (NG)	K	gas-element temperature
TG0	K	inlet gas temperature
THTA	rad	azimuthal angle
TMASSF	kg/m · sec^{-1}	total mass inflow
TN		initial value of T in Newton-Raphson method
TW (NW)	K	wall element temperature
TW1, TW2	K	wall temperature
VG	m^3	gas element volume
WP (NW)		data of wall elements transferred to subroutine PRTDAT
X0	m	X coordinate of emitting point of energy particle
X1, XW	m	X coordinate where locus of energy particle hits wall

XC		X coordinate of the center of the emitting element
XCT		X coordinate of the center of the target element
XE	m	X coordinate of exit point of energy particle from gas element [X coordinate of end point of locus of energy particle (RAT1)]
XI	m	X coordinate of incident point of energy particle to gas element
XK		absorption length
Y0	m	Y coordinate of emitting point of energy particle
Y1, YW	m	Y coordinate where locus of energy particle hits wall
YE	m	Y coordinate of exit point of energy particle from gas element [Y coordinate of end point of locus of energy particle (RAT1)]
YI	m	Y coordinate of incident point of energy particle to gas element

AUTHOR INDEX

Numbers in parentheses indicate reference numbers.

Andersson, B. A., 187 (39)
Balakrishnan, X. X., 183 (35)
Cohen, E. S., 36 (23)
Edwards, D. K., 183 (33–35)
Fujisaki, M., 181 (32), 186 (32)
Fukuchi, T., x (11), x (13, 14), 146 (28, 29), 147 (28)
Funazaki, K., x (9), 174 (9)
Funazu, M., viii (3)
Guo, K.-F., 167 (30)
Guo, K.-H., ix (15), 187 (36–38), 193 (37, 38)
Hayasaka, H., ix (6), ix (11), 146 (28), 147 (28)
Heaslet, M. A., 27 (22)
Hottel, H. C., 14 (20), 36 (23, 25)
Howell, J. R., 23 (21), 36 (24)
Ichiraku, Y., 196 (42)
Ito, S., 196 (42)
Johnsson, F., 187 (39)
Kaneda, H., ix (15), 187 (36–38), 193 (37, 38)
Katayama, T., ix (14), 167 (30)
Kawaski, M., ix (9), 174 (9)
Kim, Y.-M., ix (17, 18), 193 (41)
Kobiyama, M., ix (5)
Komatsu, T., 181 (32), 186 (32)
Kosaka, S., 181 (32), 186 (32)
Kudo, K., ix (6, 7), ix (9–11, 13–15, 17–19), 46 (6, 7), 146 (28, 29), 147 (28), 167 (30), 174 (9), 181 (31, 32), 186 (32), 187 (36–38), 193 (37, 38, 41)
Kumagai, N., ix (14)
Kunitomo, T., 124 (27)
Kusama, A., ix (6)
Leckner, B., 187 (39)
Martin, H., 192 (40)
Matsumura, M., ix (15), 187 (36–38), 193 (37, 38), 196 (42)

Matsuoka, K., 124 (27)
Menguç, M. P., ix (12)
Mochida, A., 181 (32), 186 (32)
Nagata, T., 167 (30)
Nakamachi, I., ix (6), ix (8, 11)
Nakamura, T., ix (8, 14)
Obata, M., ix (9), 174 (9), 181 (31)
Oguma, M., ix (6)
Oguri, T., 124 (27)
Ohtaka, M., 181 (32), 186 (32)
Okigami, N., ix (6)
Omori, T., ix (8, 10)
Otaka, M., ix (19), 181 (31)
Robert, F. W., 27 (22)
Ross, S. M., 51 (26)
Saeki, T., 196 (42)
Saito, T., ix (5)
Sarofim, A. F., 36 (25)
Siegel, R., 23 (21)
Sugiyama, K., viii (4)
Sumarsono, M., 181 (31)
Sumarusono, M., ix (19)
Taniguchi, H., viii (1–7), ix (8–11, 13–15, 17–19), 46 (6, 7), 146 (28, 29), 147 (28), 167 (30), 174 (9), 181 (31, 32), 186 (32), 187 (36–38), 193 (37–41)
Taniguchi, K., viii (4)
Tien, C. L., ix (16)
Viskanta, R., ix (12)
Wang, Y., 187 (38), 193 (38)
Yang, W.-J., ix (6, 7), ix (11, 15, 17, 18), 46 (6, 7), 146 (28), 147 (28), 187 (36–38), 193 (37–41)
Yasuzawa, K., ix (8)
Zhang, Y.-Z., ix (15), 187 (37), 193 (37)

SUBJECT INDEX

A

Absorption
  by solid walls, 61–85
  gas absorption
    Monte Carlo method, 62–66, 71–74, 80–85
    radiative heat transfer, 93–95
    scattering and, 92–99
    simulation, 50–56
Absorption coefficient, *see* Gas absorption coefficient
Absorption probability, 51–52
Absorptive power, 11
Absorptivity, 1, 15
ABSORP variable, 53, 204
ABSPR variable, 53, 55, 204
Aircraft engines, combustion chambers, 173–181
Albedo, scattering, 17–18, 95, 152–154, 190
Anisotropic scattering, 146–154
Attenuation, radiation intensity, 148–149
Attenuation constant, gas-particle mixture, 17
Attenuation distance, 93, 95

B

BEER program, 53–55, 203
Beer's law, 14, 17, 34
Blackbody radiation, 7–9, 10, 11, 12, 63
Boiler furnaces, 158–167

C

Circulating fluidized bed boiler (CFBB), 187–193
Combustion chambers, jet engines, 173–181
Combustion gas, 181–187
Computer programs, 203
  BEER, 53–55
  printing instructions, 203
  RADIAN, 28–29, 46, 107–133, 158–167
  RADIAN1, 203
  RADIANW, 130–146
  RAT1, 64–75, 80, 83, 88, 122
  RAT2, 76–84, 88, 109, 122
  TFM, 27–29
  variables list, 204–208
  ZM, 39–42
Computer simulations, *see* Simulations
Convective heat transfer coefficient, 48
Convective heat transfer rate, 98
CPO variable, 161, 204
CP variable, 161, 204
Cylindrical coordinate system, radiative heat transfer, 99–102

D

DAK variable, 161, 204
Del, 19
DELTAT variable, 161, 204
Direct exchange area, 36–38
Directional absorptivity, 15
Directional emissivity, 14
Divergence, radiative heat flux equation, 20, 45
DLW variable, 161, 204
DXG variable, 161, 204
DYG variable, 161, 204

E

Electromagnetic spectrum, 6–7
Electromagnetic waves, 3–4
Emission, 4
  from gas volume, 56–59
  from solid walls, 59–61
Emissive power, thermal radiation, 10
Emissivity, 11, 12, 14
Energy balance equations, 19–20
Enthalpy
  exit gas, furnace, 157
  gas-wall system, 48

F

Flame
  absorption coefficient, 124–125
  boiler furnace, 158, 161, 164–167
  gas reformer furnace, 169–170
  jet engine combustion chamber, 177, 180–181
Flight distance, 52–53, 96
Furnaces
  boiler furnaces, 158–167
  gas-fired furnaces, 196–199
  gas reformers, 167–173
  radiative–convective heat transfer, 128, 129, 144, 164-166
  with throughflow and heat-generating region, 154–157

G

Gas absorption
  radiative heat transfer analysis, 93–95
  scattering and, 92–99
  simulation, 50–56
Gas absorption coefficient, 13–14, 52, 64, 95, 161
  in a flame, 124, 125
  monochromatic, 183
  nonuniform, 81–85
  uniform, 62–81
Gas elements
  heat balance equations, 47
  simulating radiative heat transfer
    absorption, 50–56, 62–66, 71–74, 80–85
    emission, 56–59
Gas-fired furnaces, 196–199
Gas–particle mixture radiation
  attenuation constant, 17
  scattering, 93–99
  scattering albedo, 17–18
  scattering phase function, 18
Gas radiation, 13
  absorption coefficient, 13–14, 52, 64, 95, 161
    in a flame, 124, 125
    monochromatic, 183
    nonuniform, 81–85
    uniform, 62–81
  directional absorptivity, 15
  directional emissivity, 14
  from isothermal gas volume, 15–16
  gas volume and solid walls
    heat balance equations, 46–49
    Monte Carlo simulation, 49–85
    READ method, 86–90, 107–146
  gray gas, 13
Gas reformer furnace, 167–173
Gas–wall system
  heat balance equations, 46–49
  Monte Carlo simulations
    emission from gas volume, 56–59
    emission from solid walls, 59–61
    gas absorption, 56–59
    reflection and absorption by solid walls, 61–85
  READ method, 86–90, 107–146
GMF systems, 161, 205
Gray gas, 13
Gray surface, 11, 12, 61, 147

H

Heat balance
  equations, 46–49, 97–98
  Monte Carlo method, 151–154
Heat flux, 25
  wall heat flux, 98, 143, 152, 154, 157
Heat transfer, *see* Radiative heat transfer

I

INDFL variable, 127, 129, 205
INDFUL variable, 129, 205
INDGW variable, 125, 161, 205
INDNT variables, 126, 161, 205
INDRDP variable, 129, 205
Industrial gas-fired furnaces, 196–199
INDWBC variable, 126–127, 161, 163, 205
Inverse transformation method, 51, 60
ISG variable, 102–103
Isotropic scattering, 157
ISW variable, 102–103
IW variable, 109, 205

J

Jet engines, combustion chambers, 173–181

SUBJECT INDEX

K
Kirchhoff's law, 11–12

L
Lambert's cosine law, 9
Light, 3–7

M
Milne–Eddington approximation, 23
Monochromatic absorption coefficient, 183
Monochromatic emissivity, 11
Monte Carlo method
  Beer's law, 53–56
  energy method, 86, 87
  heat balance, 151–154
  industrial applications
    boiler furnaces, 158–167
    circulating fluidized bed boiler, 187–193
    gas reformer furnaces, 167–173
    jet engine combustion chambers, 173–181
    nongray gas layer, 181–187
    three-dimensional systems, 193–199
  photons, 56–57
  radiative heat transfer
    emission, 56–61
    gas absorption, 50–56
    reflection and absorption by solid walls, 61–85
  RAT1 program, 64–75, 80, 83, 88, 122, 203
  RAT2 program, 76–84, 88, 109, 122, 203
  READ method, 46, 86–91, 107–157

N
NGM variable, 124, 206
ngprnt function, 129
Nongray gas layer, 181–187
Nonorthogonal boundary, radiative heat transfer, 99–104
NRAY variable, 53, 62–63, 66, 88–89, 96–97, 127, 206
NWM variable, 124, 206
NXTGAS subroutine, 109

P
Packed spheres, 193–195
Photons, Monte Carlo method, 45–46
Planck's law, 7–8
PRTDAT subroutine, 122, 123, 127, 162

Q
QG variable, 161, 207
QND variable, 27, 207
Quanta, definition, 3, 4
Quantum theory, 4–5
QW values, 163, 201

R
RADIAN program, 28–29, 46, 203
  absorbing–emitting gas, 107–109, 122–130
  boiler furnaces, analysis, 158–167
  output, 130–133
  program listing, 110–122
RADIAN1 program, 203
RADIANW program, 203
  output, 143–144
  program listing, 134–142
  surfaces separated by nonparticipating gas, 130, 132, 145–146
Radiation, *see also* Gas-particle mixture radiation; Gas radiation; Radiative–convective system; Radiative heat transfer; Thermal radiation
  blackbody, 7–9
  definition, 3, 7
  emissive power, 10
  Kirchhoff's law, 11–12
  Lambert's cosine law, 9
  Planck's law, 7–8
  scattering, 17
    anisotropic, 146–154
    isotropic, 157
    radiative–convective system, 92–99
  solid surfaces, 10–12
  spectroradiometric curves, 7–9
  Stefan–Boltzmann law, 8–9
  Wein's displacement law, 8
Radiation absorption, *see* Absorption
Radiation emission, *see* Emission
Radiation energy absorption distribution method, *see* READ method
Radiation intensity
  attenuation, 148–149
  source function, 21–23
  thermal radiation, 9–10, 13–14

SUBJECT INDEX

Radiative–convective system, 45–46
  analysis
    energy method, 86, 87
    READ method, 46, 86–90, 107–146
    heat balance equations, 46–49
    heat transfer in furnaces, 128, 129, 144, 164–166
    scattering by particles, 93–99
  Monte Carlo simulation, 49–85
Radiative energy
  absorption–reflection characteristics, 61–85
  computation, 86, 87
  gas layer, 50–53
  gas–wall system, 47–48, 60
  Monte Carlo simulations, 50
  READ method, 46, 86–90
    absorbing–emitting gas, 146–157
    nonparticipating gas, 130–146
    RADIAN, 107–133
Radiative heat flux equation, divergence, 20
Radiative heat flux vector, 20, 25, 98
Radiative heat transfer
  absorbing–emitting gas
    RADIAN, 107–133
    READ, 146–157
  cylindrical coordinate system, 99–102
  gas volume and solid walls, 46–49
  heat balance equations, 46–49
  industrial applications
    boiler furnaces, 158–167
    circulating fluidized bed boiler, 187–193
    gas reformer furnaces, 167–173
    jet engine combustion chambers, 173–181
    nongray gas layer, 181–187
    three-dimensional systems, 193–199
  Monte Carlo method
    emission, 56–61
    gas absorption, 50–56
    READ method, 46, 84–91, 107–157
    reflection and absorption by solid walls, 61–85
  nonorthogonal boundary, 99–104
  nonparticipating gas, 133–146
  scattering, 92–99
  surfaces separated by nonparticipating gas, 130–146
Radiative heat transfer equations, 19–23
Radiosity, 34–35

Radio waves, 3
RANDOM subroutine, 53, 109–110
RAN variable, 53, 207
RAT1 program, 64–75, 80, 83, 88, 203
  flowchart, 72
  output, 73–75
  program listing, 67–71
RAT2 program, 81–84, 88, 109, 203
  flowchart, 82
  output, 83–84
  program listing, 76–80
READC, 109
READ method, 46, 86–91
  absorbing–emitting gas, 146–157
  nonparticipating gas, 130–148
  RADIAN, 107–133
READ values, 72–73, 84–85, 88, 89, 109
  cylindrical coordinate system, 100
  scattering phenomena, 96–97
  stopping printout, 129–130
Reflection, by solid walls, 61–85
Reflective power, 11
Reflectivity, definition, 11

S

Scattering, 17
  anisotropic, 146–154
  isotropic, 157
  radiative–convective system, 92–99
Scattering albedo, 17–18, 95, 152–154, 190
Scattering coefficient, 149–150
Scattering phase function, gas–particle mixture, 18
Schuster–Schwarzschild approximation, 23–31
Self-absorption, 48, 58, 96–97, 131
Simulations
  BEER program, 53–55
  radiative heat transfer, 62–85
    gas absorption, 50–56
    radiation emission from gas volume, 56–59
    radiation emission from solid walls, 59–61
    reflection and absorption by solid walls, 61–85
  RAT1 program, 64–75, 80, 83, 88
  RAT2 program, 76–84, 88, 109
SMIN variable, 81–82, 207

SUBJECT INDEX

Solid surfaces, *see also* Wall elements
    Kirchhoff's law, 11–12
    thermal radiation, 10–12
Source function, radiation intensity, 21–23
Spectroradiometric curves, thermal radiation, 7–9
Stefan–Boltzmann law, 8–9

T

TFM program, 27–29, 203
Thermal radiation, 9–14; *see also* entries under Radiation and Radiative
Three-dimensional systems
    industrial gas-fired furnaces, 196–199
    packed spheres, 193–195
Total absorption coefficient, 149
Total radiant energy, from isothermal gas volume, 15–16
Transmissive power, 11
Transmissivity, 11
Transport equations, 21
    solutions
        two flux method, 23–31
        zone method, 31–42

W

Wall elements, 10–12, 124, 147–148
    heat balance equations, 47
    radiative energy, 48
    simulating radiative heat transfer
        absorption, 74–75, 80
        emission from solid walls, 59–61
        reflection and absorption, 61–62
Wall heat flux, 98, 143, 152, 154, 157
Wave mechanics, 3, 5–6
Wave velocity, 4
Wein's displacement law, 8
wgprnt function, 129

X

X-rays, 3

Z

ZM program, 39–42, 203
Zone method, 31–42

ISBN 0-12-020027-9